SpringerBriefs in Applied Sciences and Technology

More information about this series at http://www.springer.com/series/8884

Filipe Manuel Clemente

Small-Sided and Conditioned Games in Soccer Training

The Science and Practical Applications

 Springer

Filipe Manuel Clemente
Instituto Politécnico de Viana do Castelo
Escola Superior de Desporto e Lazer
Melgaço
Portugal

ISSN 2191-530X   ISSN 2191-5318 (electronic)
SpringerBriefs in Applied Sciences and Technology
ISBN 978-981-10-0879-5   ISBN 978-981-10-0880-1 (eBook)
DOI 10.1007/978-981-10-0880-1

Library of Congress Control Number: 2016935964

Printed on acid-free paper

This Springer imprint is published by Springer Nature
The registered company is Springer Science+Business Media Singapore Pte Ltd.

Acknowledgments

The author would like to thank Instituto Politécnico de Viana do Castelo—Escola superior de Melgaço and Instituto de Telecomunicações—Covilhã for the institutional support to make this book.

The author would also to thank to Prof. Fernando Manuel Lourenço Martins and Rui Sousa Mendes for their scientific contribution in previous works and for their suggestions for this book.

Finally, the author would like to thank his family for their permanent support and patience with his scientific activity. For this reason, this book is dedicated to his family.

This work was supported by the FCT project PEst-OE/EEI/LA0008/2013.

Contents

About the Author

Filipe Manuel Clemente is Professor at Instituto Politécnico de Viana do Castelo, Escola Superior de Desporto e Lazer (Portugal) and researcher in Instituto de Telecomunicações, Delegação da Covilhã (Portugal). He has a Ph.D. in Sport Sciences—Sports training from Faculty of Sport Sciences and Physical Education in University of Coimbra (Portugal). His research in sports training and sports medicine has led to more than 110 publications. He has conducted studies in computational tactical metrics, network analysis applied to team sports analysis, small-sided and conditioned games, physical activity and health and sports medicine. He is guest editor of Sports Performance and Exercise collection in Springer Plus journal. He is also editor in three other scientific journals and reviewer in nine journals indexed on Web of Knowledge.
For further details see: http://www.researchgate.net/profile/Filipe_Clemente
Contact: Filipe.clemente5@gmail.com

Chapter 1
Small-Sided and Conditioned Games: An Integrative Training Approach

Abstract The specificity of the training requires that the soccer tasks improve all performance indicators associated with the game. For that reason, both technical and tactical indicators must be integrated in tasks that are typically used to develop physiological and physical variables. Therefore, small-sided and conditioned games (SSCG) have been recommended as the specific tasks that are required to apply in soccer training. Based on that, this chapter aims to provide relevant information that justifies the use of these smaller and adjusted versions of the formal game in the context of the training.

Keywords Small-sided and conditioned games · Decision-making · Tactics · Technical · Soccer · Football · Sports training

1.1 Should We Call Small-Sided Games (SSG) or Small-Sided and Conditioned Games (SSCG)?

Small-sided games (SSG) are typically described as smaller versions of the formal game (Hill-Haas et al. 2011). These games have been very popular in the last decade (Halouani et al. 2014; Owen et al. 2004) and are mostly used to optimize the time of training and the physiological/physical capabilities by following the main principle of the training methodology: the specificity (Clemente et al. 2014a).

SSGs are commonly smaller versions of the game that adjust the number of players (format) and the size of the field (Reilly and White 2004). These adjustments make possible to increase the individual participation of the players in the game and also increases the acute physiological responses (Castellano et al. 2013; Dellal et al. 2012).

The first studies in SSGs were mainly focused on the acute physiological effects promoted by changing the format and the size of the field (Aroso et al. 2004; Katis and Kellis 2009; Owen et al. 2004). Nevertheless, coaches often change some rules of play and even the structure of the game during training sessions (Davids et al.

F.M. Clemente, *Small-Sided and Conditioned Games in Soccer Training*,
SpringerBriefs in Applied Sciences and Technology,
DOI 10.1007/978-981-10-0880-1_1

2013). These significant adjustments use some task conditions to augment the perception of players for specific topics, mostly tactical issues (Clemente et al. 2014b). Based on that, these games are no more only smaller versions of the game but new versions of the game, thus the concept small-sided game may not characterize the full concept of these variations.

Trying to provide a new understanding of the adjusted versions of the game, the concept of small-sided and conditioned games (SSCG) have been recently used (Clemente et al. 2014c; Davids et al. 2013). The concept of SCCG is more in-depth than the SSG. The typical smaller versions of the game (SSGs) are often used for fitness development and used as an alternative to the traditional running-based activities (Dellal et al. 2008; Hill-Haas et al. 2009). The main focus is to provide an extra motivation to players to run in high intensities. In other hand, the SCCGs can be used to help learners or expert players to gain more experience in picking up specifying information for continuously regulating interpersonal interactions with teammates and opponents during the match (Davids et al. 2013). These conditioned games can be developed to augment the perception of the players for specific tactical topics and also to develop fitness and conditioning (Clemente et al. 2014a, b, c). For that reason, the tactical complexity of SSCGs can be more rich and fruitful than regular SSGs. Nevertheless, in our perspective both concepts can be integrated as one-single concept: small-sided and conditioned games (SSCG). For that reason, in this book the concept of SSCG can be associated with regular smaller versions or with more complex and adjusted versions of the game.

The main focus of our book is the application of SSCGs on soccer training. Nevertheless, these conditioned games are often used in other team sports. In fact, the pertinence of these games for coaches has been researched (Clemente et al. 2015; Leite et al. 2009; Serrano et al. 2013; Siokos 2011). In the case of basketball it was asked to 185 coaches which kind of tasks (SSCGs, offensive superiority games, defensive superiority games, formal game, offense and defense) must be recommended for specific stage of expertise (initiation, orientation, specialization, and high performance) (Leite et al. 2009). The results revealed that SSCGs, formal game, and offense are important tasks to use in any stage of expertise, thus suggesting the great pertinence of the conditioned games for the training. In other hand, in futsal (indoor soccer) a similar study revealed that coaches frequently used more SSCGs in the context of elite players, progressively increasing from the novice stages (Serrano et al. 2013).

Nevertheless, the use of SSCGs depends from the coaches' level and from their background. The adjusted versions of the game require that coaches really know the game and the didactics to modify a particular variable that augment the perception of players for the specific tactical topic. A study conducted in Australia disclosures that coaches are not sufficiently qualified to coach by using SSCGs (Siokos 2011). In this study it was revealed that coaches that worked in under-6 and under-8 opted for too much free-play style, without a technical or tactical orientation/topic. For that reason, it was suggested that coaches need to have a conceptual understanding of how to best utilize SSCGs in a supportive and inclusive coaching environment (Siokos 2011). Nevertheless, in expert coaches it was verified their better capabilities to correctly apply SSCGs in the context of soccer training (Clemente et al. 2015). In a study that was asked to an expert coach to develop SSCGs to match with specific heart rate intensities, it was possible to verify that the application of the designed tasks resulted in moderate-to-strength correlations with the heart rate responses expected by the coach (Clemente et al. 2015).

Some elite coaches have also been emphasizing the importance of using drill-based tasks that emulate the dynamic of the game. One of the main coaches that advocate for the use of SSCGs as the main tasks to develop his teams is José Mourinho. During a long interview about their training methodologies, Rui Faria (member of the Mourinho's staff) said (Oliveira et al. 2006):

> The ultimate goal is to play. For that reason, sports training only can mean one thing: doing at playing. If the goal is to improve the quality of the game and the organization, these parameters only can be achieved by using training situations where such organization can be developed.

Following the concept of an integrative approach that can be reached by using SSCGs, Mourinho also said (Figueroa and Mourão 2009):

> The beauty of this type of training [drill-based tasks] it is the possibility to develop at the same time many things. It is hard to define the goal of this task [a specific SSCGs that Mourinho applied] because he is very rich.

SSCGs are adjusted versions of the game that improve the possibility to learn or to develop a tactical topic and at the same time may also provide the opportunity to develop the fitness (Halouani et al. 2014). Their opportunities are evident in elite training and also in recreational mode and for that reason it is important to highlight the main benefits of these games for the soccer training (Krustrup et al. 2010). The justifications to adhere by these tasks can be briefly identified in the following section.

1.2 Why Should We Use Small-Sided and Conditioned Games?

The traditional approach to soccer training was to incorporate the fitness training within the overall session but without the ball when doing so, thus making such training a running-based activity (Reilly 2005). Such approach was adopted based on the traditional training methodology and also considering the beginnings of the training periodization (Turner 2011). The use of such approach focused fitness in weight training and running and only in a second part of the training comprised ball skills, training drills, and games (Reilly 2005). This separation between fitness and technical/tactical training not accomplished one of the main principles of the training: the specificity.

The specificity cannot be only understood as training responses/adaptations that are tightly coupled to the mode, frequency, and duration of exercise performed

(Hawley 2008). This physiological point of view may reduce the big-picture of the fundamental concept. Running-based activities may emulate the physiological conditions of soccer, nevertheless it lacks something to be the soccer: the game, their dynamic, and most important the ball (Davids et al. 2005; Gréhaigne et al. 1997).

The analytical and traditional approach advocate that the fitness training must be focused in running-based activities and weight training. Such approach frequently leads to longer training sessions because the first part of the training is totally different from the second part (Reilly 2005). In this methodological option, the fitness coach does not participate in the second part of the training and the head coach does not participate in the first part of the training. For that reason the synchronization between staff can be more difficult. In summary, the 'classical' approach to the fitness training soccer leads to longer sessions and also to a separation between physical and the technical/tactical development.

In traditional training, tasks often are designed for performance without opposition or with passive opposition to simplify decision-making during repetitive drill practice with small levels of variability (Davids et al. 2013). These tasks are designed to increase the accuracy and precision of the actions and for that reason the ecological validity is reduced. Typically, these traditional drills can be characterized and easily identified by the presence of static markers to benefit repetition of discrete performance (Chow et al. 2007).

If the smaller complexity can be understood as a benefit to guarantee the greater performance in a specific task, it is also important to highlight that such small complexity may be not adequate with the specificity of the soccer. The dynamic of soccer requires a permanent intra- and inter-synchronization with teammates and opponents and, for that reason, tactics are always present even in moments without the ball (Duarte et al. 2012; Vilar et al. 2014). Such tactical improvement depends from the capacity to be fast into 'read' the dynamic of the game (González-Víllora et al. 2011). Nevertheless, if a half of the training session occurs in isolated and fragmented tasks without the essence of the game (the cooperation–opposition relationship), it is half of the training that players are not improving their abilities to quickly analyze the contextual variables and make decisions about them (McGarry 2005).

On the other hand, by using drill-based activities it is possible to include fitness workout and also the development of decision-making situations in conditioned games (Serra-Olivares et al. 2015). This option by drill-based activities guarantees the simulation of movement patterns of team sport, while maintaining a competitive environment where players must perform under pressure and while fatigue (Gabbett 2008). More important than that, drill-based activities provide an additional challenge to team-sport players that would not normally be present in non-skill tasks associated with conditioning (Gabbett et al. 2009).

Obviously that the use of drill-based activities (SSG, SSCG) may lead to an increase in the variability of acute physiological effects (Hill-Haas et al. 2009). In the same task, the individual participation of each player is different and the specific variability of the game may induce different results in the players. However, evidence suggest that a coach can attempt to control the intensity of soccer drills by designing specific drills and by manipulating parameters that may increase or reduce the training load of the task (Little and Williams 2007; Little 2009).

There is other specific critic that is currently used against the use of drill-based activities in soccer training context: the adaptations are not similar in comparison with running-based activities. However, the scientific evidences can easily refute such assumption (Delextrat and Martínez 2014; Impellizzeri et al. 2006b; Reilly and White 2004). The comparisons between training programs based on SSCGs and High-Intensity Interval Training (HIIT) revealed similar adaptations in aerobic, anaerobic, speed, and power (Delextrat and Martínez 2014; Radziminski et al. 2013). Similar findings were observed in acute physiological effects of SSCGs and running-based activities (Dellal et al. 2008; Radziminski et al. 2013).

In summary, the use of drill-based activities induces similar effects in fitness development than running-based activities. Moreover, drill-based activities also improve the development of tactical thinking, the ability to make decisions during activities and the capacity to increase the inter-synchronization with teammates and with the opponents. The scientific evidences also highlight the possibility to use SCCGs in any competitive level and learning stages, thus being appropriate to be used in any context. Finally, the use of SSCGs is associated with high motivation levels and engagement of the players, thus promoting benefits in the individual commitment with the training sessions and the activities promoted by the coach.

After the summary that justifies the use of SSCGs in training context there is one doubt that readers may have at this time: how should I use SSCGs in my training context? One of the main purposes of this book is to answer to this question. Nevertheless, the next subsection will try to briefly provide some lights about this particular and reasonable question.

1.3 How Should We Use Small-Sided and Conditioned Games?

The use of SSCGs must follow a multidisciplinary approach that integrates knowledge of physiology, strength and conditioning, training methodology, pedagogy, didactics, and phycology. In fact, the use of SSCGs requires a specific training perspective that depends from the context, the knowhow of the staff and the conditions.

To properly use SSCGs in training context it is required a holistic approach that attempts to engage players on physical, cognitive, and emotional levels (Renshaw et al. 2015). SSCGs may be used to learning situations or to elite players. The approach required for both must be necessarily different. In the case of learning context, the constraints-led approach (CLA) may offer appropriate solutions that will help to increase the possibilities to self-organize the learning systems based on the conditions or constraints put in place by the coach (Lee et al. 2014; Renshaw et al. 2015). The main advantage of CLA is to design SSCGs to help learners to self-organize the performance and augment the perception for key points in game dynamics, thus improving their ability to anticipate the actions and adjust their behavior based on the variability of the context (Chow et al. 2006; Davids et al. 2006).

In learning stages it is also important to consider that specific teaching style may fit better than others in the process of using SSCGs (Mosston and Ashworth 2002). Briefly, there are tow main dimensions of teaching styles: (i) reproductive teaching styles; and (ii) productive teaching styles. The reproductive teaching styles generally are associated with a leadership based on command or in very controlled situations (Kingston et al. 2005). on the other hand, reproductive styles such as guided discovery style are better to use in the approach to the SSCGs (Brooker et al. 2000). In the case of reproductive style such as discovery style the learning process is focused on the player and the task must be designed to benefit the understanding of key points of the game (Webb and Pearson 2008). Such approach forces the player to solve the tactical problem that was designed by the coach.

Therefore, the use of SSCGs cannot be limited to the use of the games in training context. A more holistic approach should be used. Such approach requires the use of a proper leadership (more tendencies to be democratic) and a strong knowhow about the game and the tactical topics that can emerge from specific task conditions (Araújo et al. 2004). If it is logic that in learning stages the coaches should control all the knowledge about the process of designing tasks to support the learning of specific skills or tactical knowledge, it is also important that in elite context the coach may understand the best variables to increase the potential for the players. The use of well-designed tasks may increase the learning of specific tactical principles or model of play that the coach wants to implement in the regular style of play of the team. Moreover, the use of SSCGs with specific tactical topics based on coaches' perspective will also contribute for a better understanding of the routines of the team.

Design of SSCGs may also be another important topic to be highlighted in this section. Coaches may design their tasks based on specific pedagogical principles, even if the ultimate goal is the elite training. For principles can be followed during the design of SSCGs (Tan et al. 2012): (i) sampling; (ii) task complexity; (iii) representation; and (iv) exaggeration. The sampling principle augments the potential of a task for similar situations in other contexts or sports. The aim of this principle is to increase the transfer of learning (Mitchell et al. 2006). Similar sports such as soccer, futsal, basketball, or handball may have some similar tactical situations that can be learned in the same task. The ability to create space to receive the ball or to create a line of pass can be generalized in team sports of invasion and for that reason a tasks specifically designed to develop such ability in soccer can be also an important contribution to augment the capacity in other contexts (Araújo et al. 2004). This is a principle that is more useful in younger and novice players than in specialization stages.

Tactical complexity is another pedagogical principle that is truly important in designing SSCGs. This principles aim to adjust the tactical complexity to the capacity of the players. In learning stages it is expected to start with simpler games if the main goal is that novice players understand the games they play (Tan et al. 2012). Thus, games with less tactical complexity must be taught first before games with greater complexity (Mitchell et al. 2006). This principle is not exclusive in learning stages. The same reality can be verified during a season in high-competitive teams. In the beginning of the season it is logic that SSCGs are less complex than in middle season, because in initial stages it is important to build the style of play in basic tactical principles selected by the coach. Moreover, in the beginning of the season there are new players that are not integrated and for that reason less complex tasks will help to find the best connections with their teammates.

Representation involves developing SCCGs for specific tactical issues that occur in particular moments of the game. The aim of this principle is to extract a specific tactical topic of the game and design a task to increase the situations in which occurs the tactical issue. This will help the players to 'repeat without repeat' (Tan et al. 2012). The players will be exposed to the same tactical issue; nevertheless the variability of the game will increase their creativity to solve the problem in different ways. For that reason, the problem occurs many times but the capacity to solve the problem will be different based on the dynamic that occurs in SSCGs.

Finally, exaggeration principle modifies some rules of the game to augment the perception of the players for a specific tactical topic. An example is to use specific spaces of the field (wings) without possibility of opponent entering to increase the participation of lateral defenders of wing midfielders and to augment the perception of the players to pass the ball for the wings. This exaggeration that not occurs in formal game will help the players to repeat specific collective patterns. Other examples of exaggeration can be the use of smaller goals or not used goals. Each modification of the official rule will promote specific adaptations and will increase the augment of the perception to do specific actions and behaviors.

In summary, the design of the task can be crucial to improve the experience and benefits of using SSCGs in training sessions. Moreover, the style of leadership and the holist approach to the problem will be also important in the moment of using SSCGs in training context. For that reason, use of SSCGs is more than use of some modified tasks in the training. Each SSCG should be properly used and such game must follow the logic of model of the play of the team. This means that a very good and successful SSCG used in a team can be the worse SSCG in other team. Based on that, this book will not provide a long list of possible SSCG. In other way, we hope to increase the capacity of the coach to design specific tasks based on their ideas for the team and for the training session.

1.4 Why Should We Research the Effects of Small-Sided and Conditioned Games?

Drill-based games are not new in soccer training. Such games come from early stages of the training and for that reason it was not science that builds the SSCGs. Nevertheless, without science would be hardest to find the specific changes that each variable promotes in players. External and internal training load must be accurately measured to identify the effects of each condition in acute responses of the players. Moreover, without accurate measures and some scientific point of view some myths may remain in the professional environment.

As described before, some coaches believe that running-based activities are more effective than SSCGs to develop fitness levels. Such skepticism can only be faced with facts and data. Conducting some research in practical context can easily generate such data and facts. Therefore, science will help in practice by summarizing the pros and cons of using SSCGs.

The beginning stage of research conducted in SSCGs analyzed the heart rate responses, perceived effort, and blood lactate concentrations after specific conditions (Aroso et al. 2004; Owen et al. 2004). After a period of analysis carried out in isolated games, some researchers decided to compare the acute effects of SSCGs and running-based activities (Dellal et al. 2012; Sassi et al. 2004). The focus on acute effects leads to a lot of information provided in the last decade. For that reason from 2011 some reviews and meta-analysis were carried out by researchers trying to help the tasks of synthetizing the information (Aguiar et al. 2012; Clemente et al. 2014; Halouani et al. 2014; Hill-Haas et al. 2011; Little 2009).

Nevertheless, the acute effects are not enough to validate the use of SSCGs instead of running-based activities. Following such idea, some researches have been conducted comparing the adaptations promoted by SSCGs and running-based programs in fitness parameters (Impellizzeri et al. 2006a; Reilly and White 2004). Moreover, the development of specific technological devices helped to track the activity profile of players during SSCGs (Aguiar et al. 2015; Casamichana and Castellano 2010).

More recently, these technological tracking systems have been helping to characterize some patterns of collective organization using specific algorithms (Frencken et al. 2011; Silva et al. 2014). The decision-making that occurs during SSCGs have been also analyzed by using some observational systems (González-Víllora et al. 2015; Serra-Olivares et al. 2015).

The contributions of these researchers have been helping to support the concept of SSCGs and their applications in soccer training. Nevertheless, there is a lot of work to do in the future and for that reason it is predictable that the next decade brings some new findings based on the effects of these games for the players. Nevertheless, by now we think it is time to summarize the most important evidences in a single space: this book. Based on that, we will try to bring to the readers the opportunity to verify the most important findings about SSCGs and discuss the implications of such results for the practice.

1.5   What You Can Expect from This Book? Why You Should Continue Reading This Book?

Both questions are fundamental for the readers. After opening the mind for the usefulness of SSCGs in the training context, it is now important to identify what is expectable to find in this book. The main commitment that the author can make with the reader is that the all information comes from scientific evidences. Well-conducted studies about SSCGs will be presented and discussed trying to justify the importance of these games in soccer training. After that, we aim that readers are able to identify the benefits and cons of SSCGs.

In the beginning of this book it will be possible to identify the reality of the game. A set of well-conducted studies will be the support of this chapter, particularly characterizing the physiological demands and time–motion profiles that occur in youth and elite competitive soccer. This characterization will make easier to identify what is need to be developed in the context of soccer training and what variables must be improved to correspond to the demands of the match. Briefly the main purpose is to justify why specificity is so important in soccer training.

After knowing the demands of the match, it is important to provide a valuable summary of the studies that compared the acute effects and the adaptations between drill-based activities (mostly SSCGs) and the running-based activities. In this chapter it is expected to discuss the variables that have been improved with structured training programs based on two different methodologies. The adaptations induced by both type of training programs in aerobic, anaerobic, power and strength, agility, and speed will be presented. This will help the reader to identify that SSCGs are valid tasks to fitness development and for that reason the question that remains is: which variables must I use to induce specific patterns of training load?

Trying to help the task of the reader, the chapters four, five, six, and seven will show the influence of different variables in acute responses. Usually, format of the game and size of the field are the variable more used by coaches. Based on that, it will have a dedicated chapter for each variable. The effects on training load (hear rate responses, blood lactate concentration, perceived exertion, time–motion profile, technical actions, and tactical behaviors) will be presented based on the scientific studies that have been conducted about both variables. Nevertheless, as described above, other variables have been used by coaches in daily training activities such as limits of touches on the ball, use of different goals or targets, the use of floated elements (neutral players), or the use of encouragement. Therefore, the chapter six will analyze the effects of these variables on performance responses of players.

After identifying the different effects of variables, it will be finally important to make the following question: How it is possible to organize the weekly periodization based on drill-based activities? Trying to help the reader to answer to that question, the chapter seven will discuss some possibilities to organize and distribute the training load by using a periodization based on drill-based activities. In this chapter it will be possible to verify some examples for micro-cycles with one and two weekly competitions and also to identify the most appropriate tasks for each day of the week. This final chapter of this book will also present the conclusions, and most important, the future directions and practical applications for soccer training. It is important to highlight that the training prescription is not an exact science and based on this it is possible to assume that other training approaches will replace this training methodology in the next years/decades. Moreover, it is also assumed that variability of contexts may influence the benefits of training methodologies based on SSCGs. Therefore, this book cannot be assumed as a manual. It is more adequate to look for this book as a guide of orientations for those who want to use drill-based activities in soccer training.

Finally, the aim of this book is not to provide a list of tasks. It is important to make this clear: our aim is to discuss the scientific evidences about SSCGs and not to provide a list of formulas and magic tasks that make the science training a fast food store. We want that the reader to use the scientific evidence to design their own tasks thus following the main principles of the training: the specificity and the individualization. To follow these principles, it is important that each coach designs their own tasks for their specific team, thus making each task a singular and unrepeatable moment of fitness development and tactical evolution.

After clearing the relationship between the author and the reader and after identifying what is expected to find in this book, it is the moment to make the final and most important question: should move on to the next chapter. We hope so!

References

Aguiar, M., Botelho, G., Lago, C., Maças, V., & Sampaio, J. (2012). A review on the effects of soccer small-sided games. *Journal of Human Kinetics, 33*, 103–113.

Aguiar, M., Gonçalves, B., Botelho, G., Lemmink, K., & Sampaio, J. (2015). Footballers' movement behaviour during 2-, 3-, 4- and 5-a-side small-sided games. *Journal of Sports Sciences*, 1–8.

Araújo, D., Davids, K., Bennett, S., Button, C., & Chapman, G. (2004). Emergence of Sport Skills under Constraints. In A. M. Williams & N. J. Hodges (Eds.), *Skill acquisition in sport: Research, theory and practice* (pp. 409–434). London, UK: Routledge, Taylor & Francis.

Aroso, J., Rebelo, A. N., & Gomes-Pereira, J. (2004). Physiological impact of selected game-related exercises. *Journal of Sports Sciences, 22*, 522.

Brooker, R., Kirk, D., Braiuka, S., & Bransgrove, A. (2000). Implementing a game sense approach to teaching junior high school basketball in a Naturalistic Setting. *European Physical Education Review, 6*, 7–26.

Casamichana, D., & Castellano, J. (2010). Time–motion, heart rate, perceptual and motor behaviour demands in small-sides soccer games: Effects of pitch size. *Journal of Sports Sciences, 28*(14), 1615–1623.

Castellano, J., Casamichana, D., & Dellal, A. (2013). Influence of game format and number of players on heart rate responses and physical demands in small-sided soccer games. *Journal of Strength & Conditioning Research, 27*, 1295–1303.

Chow, J. Y., Davids, K., Button, C., & Koh, M. (2007). Variation in coordination of a discrete multiarticular action as a function of skill level. *Journal of Motor Behavior, 39*(6), 463–479. doi:10.3200/JMBR.39.6.463-480.

Chow, J. Y., Davids, K., Button, C., Shuttleworth, R., Renshaw, I., & Araujo, D. (2006). Nonlinear pedagogy: A constraints-led framework for understanding emergence of game play and movement skills. *Nonlinear Dynamics, Psychology, and Life Sciences, 10*(1), 71–103.

Clemente, F. M., Lourenço, F. M., & Mendes, R. S. (2014a). Developing aerobic and anaerobic fitness using small-sided soccer games: Methodological proposals. *Strength and Conditioning Journal, 36*(3), 76–87.

Clemente, F. M., Martins, F. M. L., & Mendes, R. S. (2014b). Periodization based on small-sided soccer games. *Strength and Conditioning Journal, 36*(5), 34–43.

Clemente, F. M., Wong, D. P., Martins, F. M. L., & Mendes, R. S. (2014c). Acute effects of the number of players and scoring method on physiological, physical, and technical performance in small-sided soccer games. *Research in Sports Medicine, 22*(4), 380–397.

Clemente, F. M., Martins, F. M. L., & Mendes, R. S. (2015). How coaches use their knowledge to develop small-sided soccer games: A case study. *South African Journal for Research in Sport, Physical Education and Recreation, 37*(1), 1–11.

Davids, K., Araújo, D., Correia, V., & Vilar, L. (2013). How small-sided and conditioned games enhance acquisition of movement and decision-making skills. *Exercise and Sport Sciences Reviews, 41*(3), 154–161.

Davids, K., Araújo, D., & Shuttleworth, R. (2005). Applications of dynamical systems theory to football. In T. Reilly, J. Cabri, & D. Araújo (Eds.), *Science and Football V* (pp. 556–569). Oxon: Routledge Taylor & Francis Group.

Davids, K., Button, C., Araújo, D., Renshaw, I., & Hristovski, R. (2006). Movement models from sports provide representative task constraints for studying adaptive behavior in human movement systems. *International Society for Adaptive Behavior, 14*(1), 73–95.

Delextrat, A., & Martínez, A. (2014). Small-sided game training improves aerobic capacity and technical skills in basketball players. *International Journal of Sports Medicine, 35*, 385–391.

Dellal, A., Chamari, K., Pintus, A., Girard, O., Cotte, T., & Keller, D. (2008). Heart rate responses during small-sided games and short intermittent running training in elite soccer players: A comparative study. *Journal of Strength and Conditioning Research, 22*(5), 1449–1457.

Dellal, A., Varliette, C., Owen, A., Chirico, E. N., & Pialoux, V. (2012). Small-sided games versus interval training in amateur soccer players: Effects on the aerobic capacity and the ability to perform intermittent exercises with changes of direction. *Journal of Strength and Conditioning Research, 26*(10), 2712–2720.

Duarte, R., Araújo, D., Correia, V., & Davids, K. (2012). Sports teams as superorganisms: Implications of sociobiological models of behaviour for research and practice in team sports performance analysis. *Sports Medicine, 42*(8), 633–642.

Figueroa, J., & Mourão, S. (2009). *Football for kids by José Mourinho*. Portugal: Modelo Continente Hipermercados S. A.

Frencken, W., Lemmink, K., Delleman, N., & Visscher, C. (2011). Oscillations of centroid position and surface area of football teams in small-sided games. *European Journal of Sport Science, 11*(4), 215–223.

Gabbett, T. (2008). Does improved decision-making ability reduce the physiological demands of game-based activities in field sport athletes? *Journal of Strength and Conditioning Research, 22*(6), 2027–2035.

Gabbett, T., Jenkins, D., & Abernethy, B. (2009). Game-based training for improving skill and physical fitness in team sport athletes. *International Journal of Sports Science & Coaching, 4*(2), 273–283.

González-Víllora, S., García-López, L. M., Pastor-Vicedo, J. C., & Contreras-Jordán, O. R. (2011). Tactical awareness and decision making in youth football players 10 years: A descriptive study. *Revista de Psicología Del Deporte, 20*(1), 79–97.

González-Víllora, S., Serra-Olivares, J., Pastor-Vicedo, J. C., & da Costa, I. T. (2015). Review of the tactical evaluation tools for youth players, assessing the tactics in team sports: football. *SpringerPlus, 4*(1), 663.

Gréhaigne, J. F., Bouthier, D., & David, B. (1997). Dynamic-system analysis of opponent relationship in collective actions in football. *Journal of Sports Sciences, 15*(2), 137–149.

Halouani, J., Chtourou, H., Gabbett, T., Chaouachi, A., & Chamari, K. (2014). Small-sided games in team sports training: Brief review. *Journal of Strength & Conditioning Research*.

Hawley, J. A. (2008). Specificity of training adaptation: Time for a rethink? *The Journal of Physiology, 586*(1), 1–2.

Hill-Haas, S. V., Coutts, A. J., Rowsell, G. J., & Dawson, B. T. (2009). Generic versus small-sided game training in soccer. *International Journal of Sports Medicine, 30*(9), 636–642.

Hill-Haas, S. V., Dawson, B., Impellizzeri, F. M., & Coutts, A. J. (2011). Physiology of small-sided games training in football. *Sports Medicine, 41*(3), 199–220.

Impellizzeri, F. M., Marcora, S. M., Castagna, C., Reilly, T., Sassi, A., Iaia, F. M., & Rampinini, E. (2006a). Physiological and performance effects of generic versus specific aerobic training in soccer players. *International Journal of Sports Medicine, 27*(6), 483–492.

Impellizzeri, F. M., Marcora, S. M., Castagna, C., Reilly, T., Sassi, A., Iaia, F. M., & Rampinini, E. (2006b). Physiological and performance effects of generic versus specific aerobic training in soccer players. *International Journal of Sports Medicine, 27*, 483–492.

Katis, A., & Kellis, E. (2009). Effects of small-sided games on physical conditioning and performance in young soccer players. *Journal of Sports Science & Medicine, 8*(3), 374.

Kingston, K., Morgan, K., & Sproule, J. (2005). Effects of different teaching styles on the teacher behaviours that influence motivational climate and pupils' motivation in physical education.

Krustrup, P., Dvorak, J., Junge, A., & Bangsbo, J. (2010). Executive summary: The health and fitness benefits of regular participation in small-sided football games. *Scandinavian Journal of Medicine & Science in Sports, 20*(Suppl 1), 132–135.

Lee, M. C. Y., Chow, J. Y., Komar, J., Tan, C. W. K., & Button, C. (2014). Nonlinear pedagogy: An effective approach to cater for individual differences in learning a sports skill. *PLoS ONE, 9* (8), e104744.

Leite, N., Vicente, P., & Sampaio, J. (2009). Coaches perceived importance of tactical items in basketball players' long term development. *Revista de Psicología Del Deporte, 18*, 481–485.

Little, T. (2009). Optimizing the use of soccer drills for physiological development. *Strength and Conditioning Journal, 31*(3), 67–74.

Little, T., & Williams, A. G. (2007). Measures of exercise intensity during soccer training drills with professional soccer players. *Journal of Strength and Conditioning Research, 21*, 367–371.

McGarry, T. (2005). Soccer as a dynamical system: Some theoretical considerations. In T. Reilly, J. Cabri, & D. Araújo (Eds.), *Science and football V* (pp. 570–579). London and New York: Routledge, Taylor & Francis Group.

Mitchell, S. A., Oslin, J. L., & Griffin, L. L. (2006). *Teaching sport concepts and skills: A tactical games approach*. Champaign, IL: Human Kinetics.

Mosston, M., & Ashworth, S. (2002). *Teaching physical education*. San Francisco, USA: Benjamin Cummins.

Oliveira, B., Amieiro, N., Resende, N., & Barreto, R. (2006). *Mourinho: Porquê tantas vitórias? [Mourinho: Why so many victories?]*. Lisboa, Portugal: Gradiva.

Owen, A., Twist, C., & Ford, P. (2004). Small-sided games: The physiological and technical effect of altering field size and player numbers. *Insight, 7*, 50–53.

Radziminski, L., Rompa, P., Barnat, W., Dargiewicz, R., & Jastrzebski, Z. (2013). A comparison of the physiological and technical effects of high-intensity running and small-sided games in young soccer players. *International Journal of Sports Science & Coaching, 8*(3), 455–465.

Reilly, T. (2005). Training specificity for soccer. *International Journal of Applied Sports Sciences, 17*(2), 17–25.

Reilly, T., & White, C. (2004). Small-sided games as an alternative to interval-training for soccer players [abstract]. *Journal of Sports Sciences, 22*(6), 559.

Renshaw, I., Araújo, D., Button, C., Chow, J. Y., Davids, K., & Moy, B. (2015). Why the constraints-led approach is not teaching games for understanding: A clarification. *Physical Education and Sport Pedagogy*, 1–22.

Sassi, R., Reilly, T., & Impellizzeri, F. M. (2004). A comparison of small-sided games and interval training in elite professional soccer players [abstract]. *Journal of Sports Sciences, 22*, 562.

Serrano, J., Shahidian, S., Sampaio, J., & Leite, N. (2013). The importance of sports performance factors and training contents from the perspective of futsal coaches. *Journal of Human Kinetics, 38*, 151–160.

Serra-Olivares, J., González-Víllora, S., García-López, L. M., & Araújo, D. (2015). Game-based approaches' pedagogical principles: Exploring task constraints in youth soccer. *Journal of Human Kinetics, 46*(1). doi:10.1515/hukin-2015-0053.

Silva, P., Duarte, R., Sampaio, J., Aguiar, P., Davids, K., Araújo, D., & Garganta, J. (2014). Field dimension and skill level constrain team tactical behaviours in small-sided and conditioned games in football. *Journal of Sports Sciences, 32*(20), 1888–1896.

Siokos, A. (2011). Determining the effectiveness of small-sided football (SSF) implementation in metropolitan association football. *International Journal of Coaching Science, 5*(1), 57–69.

Tan, C. W. K., Chow, J. Y., & Davids, K. (2012). "How does TGfU work?": Examining the relationship between learning design in TGfU and a nonlinear pedagogy. *Physical Education and Sport Pedagogy, 17*(4), 331–348.

Turner, A. (2011). The science and practice of periodization: A brief review. *Strength and Conditioning Journal, 33*(1), 34–46.

Vilar, L., Araújo, D., Davids, K., Travassos, B., Duarte, R., & Parreira, J. (2014). Interpersonal coordination tendencies supporting the creation/prevention of goal scoring opportunities in futsal. *European Journal of Sport Science, 14*(1), 28–35.

Webb, P., & Pearson, P. (2008). An integrated approach to teaching games for understanding (TGfU). *1st Asia Pacific Sport in Education Conference*, Adelaide.

Chapter 2
Physiological Demands of the Soccer and Time–Motion Profile

Abstract The training process must follow the specificity of the sport. Thus, identify the physiological demands of the game, the time–motion profile of the players, the physical specificities of the players, and the technical actions and tactical behaviors that are most common in soccer which is important. By knowing these characteristics it will be easier to develop the training tasks and correctly prescribe these tasks in the weekly periodization.

Keywords Physiological demands · Time–motion analysis · Match analysis · Soccer · Football · Sports performance

2.1 Introduction

Soccer is an invasive team sport with duration of 90 min with intermittent regimen of effort (Reilly and Williams 2003). This intermittent regimen depends from many contextual variables that determine the pace of the players and the dynamics of the game (Carling et al. 2005). The type of actions and skills and the profile of motion lead to different interventions that coach may prescribe and for that reason the overall results should be carefully analyzed by different variables such as tactical role, tactical lineup of the team, or the style of play (Carling 2013).

Usually, the motion analysis and more particularly the intensity of running and distance covered provide relevant information about the activity profile that occurs in match (Buchheit et al. 2014). The work-rate measure can be broken down into discrete actions of a player for a whole game (Rampinini et al. 2007). The activity or activities of players can be classified based on type, intensity (or quality), duration (or distance), and frequency (Reilly and Williams 2003). Nowadays, some technological devices provide the opportunity to measure such indicators (Clemente et al. 2014a): (i) multi-camera tracking systems; (ii) global positioning system (GPS); and (iii) radio-frequency identification (RFID).

Only activity does not characterize the demands of the game. For that reason, some physiological responses (heart rate, rated perceived exertion, or blood lactate

© The Author(s) 2016
F.M. Clemente, *Small-Sided and Conditioned Games in Soccer Training*,
SpringerBriefs in Applied Sciences and Technology,
DOI 10.1007/978-981-10-0880-1_2

concentrations) have been also monitored during games (Datson et al. 2014; Mohr et al. 2005). By using such information it will be possible to better understand the physical and physiological requirements of the game, thus improving the capacity to develop the specificity of the training (Little 2009; Turner and Stewart 2014).

This chapter aims to examine some aspects of the exercise intensities in soccer. Acute physiological responses, time–motion profile of players, and match analysis will be the focus of this chapter. In the end of this chapter, the practical implications for training context will be discussed.

2.2 Time–Motion Analysis

Elite soccer players commonly cover values of 10–12 km during a game (Carling et al. 2008; Di Salvo et al. 2007; Stroyer et al. 2004). The majority of studies report that central midfielders and wide defenders run the longest distances during a match and central defenders and strikers the shortest distances (excluding goalkeepers) (Clemente et al. 2013; Di Salvo et al. 2007; Mohr et al. 2003). Actually, the linking role of central midfielders may determine the greater distances covered (Mendez-Villanueva et al. 2012; Reilly 2007a, b). Defenders perform the largest amount of jogging, skipping, and shuffling movements and spend a significantly smaller amount of time sprinting and running than other players (Bloomfield et al. 2007). Another evidence is that professional and elite players run longer distances than nonprofessional or moderate players (Ekblom 1986). In this particular case, top class players may perform more 28 % of high-intensity running and 58 % of sprint than moderate players (Mohr et al. 2003). The effect of fatigue induce a decrease of 5–10 % in the total distance from the first to the second half of the match in the majority of the cases reported (Carling et al. 2005; Mohr et al. 2003; Rienzi et al. 2000).

The intermittent regimen of soccer can be associated with the evidence that a sprint bout occurs every 90 s, each lasting an average of 2–4 s (Bangsbo et al. 1991; Rienzi et al. 2000; Stølen et al. 2005). Generally, the average distance covered at high intensity is 10 % (Carling et al. 2008). Some results suggested that wide midfielders, attackers, and wide defenders covered higher total sprint distance than central defenders and central midfielders (Bradley et al. 2009; Di Salvo et al. 2007, 2010). Wide midfielders performs more sprints (>25.2 km h^{-1}), followed by attackers and wide defenders (Di Salvo et al. 2007, 2010). Central defenders perform fewer explosive and leading sprints than all other field positions (Di Salvo et al. 2010). A more recent study that analyzed the distance covered at low, moderate, and high-acceleration and deceleration revealed that on average 18 % of total distance covered is done so whilst accelerating or decelerating at a rate greater than 1 ms^{-2} (Akenhead et al. 2013). The authors also revealed that 7.5, 4.3, and 3.3 % of total distance is covered at 1–2 ms^{-2}, 2–3 ms^{-2}, and >3 ms^{-2}, respectively. It was concluded in this study that time-dependent reductions in distances covered suggest that acceleration and deceleration capability are acutely compromised during match play (Akenhead et al. 2013).

In youth levels, some studies that analyzed the sprinting performance during match play indicated that younger players competed at higher relative running intensities than their older counterparts (Buchheit et al. 2010; Mendez-Villanueva et al. 2012). By the other hand, it was possible to verify that few differences in match work rate were found between groups in a comparative study from U12 to U16 (Harley et al. 2010). Also in youth players some findings suggested that irrespective of the age group, players covered less distance in the second half than in the first half (Mendez-Villanueva et al. 2012). As similar to elite, in youth soccer generally midfielder covered the greatest distance at low relative speeds while striker displayed the lowest distance at intensities below maximal aerobic speed (Mendez-Villanueva et al. 2012).

During a match each player performs 1000–1400 short activities changing every 4–6 s (Bangsbo et al. 1991; Reilly and Thomas 1976; Stølen et al. 2005). The ratio of low intensity to high-intensity efforts is about 5:2 in terms of distance covered (Reilly and Williams 2003). Nevertheless, based on time the ratio may achieve 1:8 in 90 % of cases with an intermittent effort profile of 2.2 s/18 s (Vigne et al. 2010). On average each outfield player has a short static rest pause of only 3 s every 2 min (Reilly 2007a, b). Sprints, high-intensity running, tackles, headings, involvements with ball, or passes are the common activities that intermittently occurred during a match. Nevertheless, in professional soccer only 1.2–2.4 % of the total distance covered by players is in possession of the ball (Carling 2010; Rampinini et al. 2009). Such activity increases the physiological stress in comparison with running without ball at the same speed (Hoff et al. 2002). In a study that analyzed the specific patterns of activity in moments with ball it was found that actions are most commonly undertaken at high running speed (>10 km h^{-1}) (Carling 2010).

2.3 Acute Physiological Responses

Both intermittent activity profile of the game and the game duration contribute for the physiological stress experienced by the players. The game duration determine the mainly dependence from the aerobic metabolism (Stølen et al. 2005). The average work intensity measured by maximal heart rate (HRmax) reveals a profile of activity close to anaerobic threshold (80–90 % HRmax or 75 %VO$_{2max}$) (Hoff et al. 2002; Mohr et al. 2005). Average blood lactate concentration of 3–6 mmol l^{-1} has been verified during matches, with specific individual cases above 12 mmol l^{-1} (Bangsbo 1994; Mohr et al. 2005). The peak values may occur in man-to-man duels (Gerisch et al. 1988). These great values suggest that the anaerobic energy system can be highly taxed during intense periods of the game (Mohr et al. 2005). Nevertheless, this intensity cannot be sustained continuously under these extreme conditions, which reflect the intermittent consequences of anaerobic metabolism during soccer match (Reilly 2007a, b).

The measuring of oxygen uptake in a match was also investigated in periods of 3 min in two players (Ogushi et al. 1993). In this study it was found values of 35–38 mL/kg/min in first half and 29–30 mL/kg/min in second half which corresponding to 56–61 and 47–49 % of maximal oxygen uptake (VO_{2max}). These values obtained from Douglas bags are far from the typical values of 70–80 % VO_{2max} found by the association between heart rate and VO_2 (Bangsbo 2014; Mohr et al. 2005). Moreover, Douglas bags may have limited the participation of players in high-intensity actions such as tackles, duels, and other energy-demanding activities (Stølen et al. 2005). The association between HR and VO_2 reveals that a soccer player may achieve 45–53 mL/kg/min of oxygen uptake during a match (Stølen et al. 2005). The raise in VO_2 leads to an increase in body temperature. In the first half the muscle temperature of vastus lateralis increased from 36 °C (before warm-up) to 39.4 °C (end of first half) and 39.2 °C (end of second half) (Mohr et al. 2004). The core temperature ranges 39–40 °C during the match (Ekblom 1986; Mohr et al. 2004).

The high-intensity that occurs in a match lead to the evidence that in the great majority of time players are rarely below 65 % HRmax, thus suggesting that blood flow to the exercising leg muscles is continuously higher than at rest, which means that oxygen delivery is high (Bangsbo 2014). For that reason, the oxidative capacity of the contracting muscles may be determinant to manage the oxygen kinetics (Bangsbo 2014; Krustrup et al. 2004).

Soccer players may perform 150–250 brief intense actions during a game (Mohr et al. 2003) which indicates that the rate of anaerobic energy turnover is high during specific periods of the game (Bangsbo 2014). The short periods of very high-intensity may indicate the great capacity of creatine phosphate breakdown, which to a great extent is re-synthesized in the following low intensity exercise periods (Bangsbo 1994). Nevertheless, the capacity to slow down after a great effort justifies that after an intense exercise during a game the muscle biopsies revealed 75 % of the level at rest of the creatine phosphate (Krustrup et al. 2006). Despite of the important contribution of ATP-CP system for very fast and powerful actions, the glycolytic system should be also considered during the game. Periods of 5 min of high-intense exertion have been associated with blood lactate concentrations of 12–16 mmol l^{-1} (Krustrup et al. 2006; Mohr et al. 2003). The scientific evidences revealed that after such intense periods there is a 5-min period of intensity lower than the average of the match (Mohr et al. 2003). Moreover, sprint performance is reduced both temporarily during a game and at the end of a soccer game, thus low glycogen levels in individual muscle fibers may explain this evidence (Krustrup et al. 2006).

The inflammatory responses to a soccer match were also studied in elite and notelite male and female soccer players (Souglis et al. 2015). Average relative exercise intensity during the match was similar in male and female players (86.9 ± 4.3 and 85.6 ± 2.3 % HRmax, respectively) and the interleukin 6 and tumor necrosis factor alpha increased 2- to 4-fold above resting values, peaking immediately after the match. Moreover, C-reactive protein and creatine kinase peaked 24 h after the match (Souglis et al. 2015).

In summary, the tactical role and the competitive level affect the high-intensity work done in a soccer game. The intermittent profile of activity and the duration of the match influence the physiological responses of the players. During match, the heart rate may vary between 80 and 90 % HRmax, blood lactate concentration may achieve peaks of 12 mmol l^{-1}, and the VO_2 varies from 36 to 50 mL/kg/min which corresponds to 56–75 % of VO_{2max}. The short but frequent periods of high intensity suggest that the rate of creatine phosphate and glycolysis are frequently used during the game. The great use of muscle glycogen in the majority of the activities conducts to a progressive decrease in intensity and to a change to oxidation of fat by the end of the match.

2.4   Physiological and Physical Profile of Soccer Players

The physical and physiological demands of soccer were briefly described in the previous section. Nevertheless, it is also important to characterize the physiological and physical profile of the players.

Studies conducted in male soccer players reveal VO_{2max} about 50–70 mL/kg/min, whilst the goalkeepers have 50–55 mL/kg/min (Gil et al. 2007a, b; Sporis et al. 2009; Wong and Wong 2009). The anaerobic threshold is reported to be between 76.6 and 90.3 % of HRmax, which arc in line with the values found during match play (Stølen et al. 2005). In a study that tested 270 soccer players from the professional first national league of Croatia, it was found that goalkeepers and attackers are the tallest and heaviest players (Sporis et al. 2009). In the same study, it was also found that attackers were the fastest players in 5-, 10- and 20-m sprint (1.39, 2.03, and 3.28 s, respectively). Goalkeepers and attackers had best results in squat jump (46.8 and 44.2 cm, respectively) and countermovement jump (48.5 and 45.3 cm, respectively). Midfielders and defenders had the greatest results in VO_{2max} (62.3 and 59.2 mL/kg/min, respectively) (Sporis et al. 2009).

A study conducted in 241 youth Spanish players (U14 to U21) (Gil et al. 2007a, b) revealed that goalkeepers were the tallest and heaviest players; this pattern has also been reported for mature elite goalkeepers (Arnason et al. 2004). By the other hand, goalkeepers revealed the lowest performance in the endurance tests (Gil et al. 2007a, b). In this particular study conducted in Spain, forwards have the best oxygen intake and the lowest cardiac frequency in the endurance test; moreover also were the fastest players in the 30-m flat and the 30-m with turns of direction (Gil et al. 2007a, b). The nationality of the players may also contribute for different characteristics. A study carried out in elite youth Asian players revealed that these players generate less force, require longer time to reach their peak force, and have a shorter jump height than Tunisian players (Chamari et al. 2004; Wong and Wong 2009). Moreover, it was also found that Asian players are shorter and had a smaller VO_{2max} than Finnish, Tunisian, and US players (Wong and Wong 2009).

Time–motion analysis carried out in female soccer players suggests that they cover 8.5–10 km during a match with an average of 5.7–6.9 km h^{-1} (Krustrup et al.

2005; Mohr et al. 2008). Despite having these smaller values, female players show similar cardiac responses with male players (85–90 % HRmax) (Krustrup et al. 2005). A study that compared players over an 18-year period revealed that 55 mL/kg/min is the standard VO_{2max} for elite female players (Haugen et al. 2014). The range of VO_{2max} may vary from 38.6 to 57.6 mL/kg/min (Jensen and Larsson 1993; Polman et al. 2004).

2.5 Match Analysis on the Game

The match analysis based on notational techniques has helped the characterization of events of soccer. One of the main conclusions is that top teams made more shots and shots of goal than less successful teams (Armatas et al. 2009; Lago-Ballesteros and Lago-Peñas 2010). Top teams also showed better effectiveness, thus they need a lower number of shots to score a goal (Lago-Ballesteros and Lago-Peñas 2010). Moreover, playing at home may guarantee high percentages of winning (50–62 %) (Lago-Peñas and Lago-Ballesteros 2011; Pollard 2006). Moreover, home advantage also increase attack indicators, such as goal scored, total shots, shots on goal, attacking moves, box moves, crosses, assists, passes made, dribbles made, and ball possession (Lago-Peñas and Lago-Ballesteros 2011).

A study carried out in 380 matches of the Spanish soccer League indicated that ball possession can be influenced by situational variables (Lago-Peñas and Dellal 2010). In this study, losing match status was associated with an increase in ball possession in comparison with winning and drawing status (Lago-Peñas and Dellal 2010). Moreover, it was also found that playing away reduces the volume of possession of the ball in comparison with home matches. Finally, this study in Spanish teams also revealed that top-placed teams had a higher percentage of ball possession per match than the less successful teams (Lago-Peñas and Dellal 2010).

The analysis to the shots carried out in 1990 and 1994 FIFA World Cup showed that were made significantly more shots per possession at longer passing sequences than there were at shorter passing sequences for successful teams (Hughes and Franks 2005). Moreover, the authors (Hughes and Franks 2005) also revealed that the conversion ratio of shots to goals is better for direct play than for possession play. Following this evidence, a study conducted in Norway Premier League revealed that counterattacks were more effective than elaborate attacks when playing against an imbalanced defense but not against a balanced defense (Tenga et al. 2010). It was also verified that a defensive balance strategy (tight pressure, present backup, and present cover) was more effective in preventing score-box possessions than the opposite tactics of imbalanced defense (loose pressure, absent backup, and absent cover) (Tenga et al. 2010).

More recently, some studies have been using some algorithms and computational methods to characterize the collective dynamic and organization of the teams (Clemente et al. 2014b). Generally, the studies revealed that players spread their positions during attacking moments and during defensive moments contract their collective organization (close interpersonal distances) (Bartlett et al. 2012; Clemente et al. 2013; Frencken et al. 2011; Moura et al. 2012). Moreover, teams tend to have numerical superiority in their central defensive zones and numerical disadvantage in central attacking zones (Clemente et al. 2015a; Vilar et al. 2013). During attacking moments, teammates tend to provide a greater ratio of coverage in vigilance (line of pass far from men with ball) than cover in support (close line of pass) (Clemente et al. 2014b).

The network analysis carried out during the possession of the ball has been also researched based on graph theory (Duch et al. 2010). The main results showed that most successful teams tend to be more homogeneous and dense in their passes distribution (Clemente et al. 2015b; Grund 2012). On the other hand, great heterogeneity in the passing sequences leads to worse performances (Grund 2012). Independently from tactical lineup, midfielders tend to be the most central player in the team into receive and into perform the passes, thus being the most prominent player in passing sequences (Clemente et al. 2015c; Duch et al. 2010; Peña and Touchette 2012). External defenders tend to be one of the players that most contribute into pass and forwards into received the ball (Clemente et al. 2015a).

2.6 Implications for Training

By having enough information about the physiological, physical, and technical/tactical demands of the soccer it is possible to prescribe the sports training with specificity. This principle of specificity means something more the adequate the physiological stimulus to the game. Running at 85 % HRmax is different from playing soccer game at 85 % HRmax. The muscle participation, the coordination, agility, and fundamentally the decision-making based on the capacity of perceive the environment it is very different. For that reason, training soccer and the specific capabilities of soccer players must be something more than just replies the internal load. For that reason, some studies have been comparing the traditional running methods with specific soccer drills based on the game (small-sided and conditioned games—SSCGs) (Dellal et al. 2008; Hill-Haas et al. 2009). The SSCGs on soccer training aims to ensure the fitness development and at the same time emulate the dynamic of the game.

In a study that compared friendly matches and SSCGs in semi-professional players it was found that the global indicators of workload (distance covered by minute, work: rest ratio, players workload per minute, and exertion per minute) were higher for SSCGs than for friendly matches (Casamichana et al. 2012). The authors suggested that SSCGs lead to greater intensities than friendly matches. Nevertheless, the maximum speed was greater, longer, and more frequent in friendly matches than in SSCGs (Casamichana et al. 2012).

The training with specificity requires emulate the real circumstances of the game. For that reason, SSCGs provides an important contribution for prescribe the exercise and at the same time develop tactical principles and increase the commitment of players with the training session. Nevertheless, to show that SSCGs can be equally efficient than traditional running methodologies for the fitness development it is required data. For that reason, the next chapter will summarize the studies that compared the acute responses and the adaptations that resulted from SSCGs and running-based training programs.

References

Akenhead, R., Hayes, P. R., Thompson, K. G., & French, D. (2013). Diminutions of acceleration and deceleration output during professional football match play. *Journal of Science and Medicine in Sport/Sports Medicine Australia, 16*(6), 556–561.

Armatas, V., Yiannakos, A., Zaggelidis, G., Papadopoulou, S., & Fragkos, N. (2009). Goal scoring patterns in Greek top leveled soccer matches. *Journal of Physical Education and Sport, 23*(2), 1–5.

Arnason, A., Sigurdsson, S. B., Gudmundsson, A., Holme, I., Engebretsen, L., & Bahr, R. (2004). Physical fitness, injuries, and team performance in soccer. *Medicine and Science in Sports and Exercise, 36*(2), 278–285.

Bangsbo, J. (1994). The physiology of soccer—with special reference to intense intermittent exercise. *Acta Physiologica Scandinavica, 619*(Supplementum 02), 1–155.

Bangsbo, J. (2014). Physiological demands of football. *Sports Science Exchange, 27*(125), 1–6.

Bangsbo, J., NØrregaard, L., & ThorsØ, F. (1991). Activity profile of competition football. *Canadian Journal of Sports Science, 16*, 110–116.

Bartlett, R., Button, C., Robins, M., Dutt-Mazumder, A., & Kennedy, G. (2012). Analysing team coordination patterns from player movement trajectories in football: Methodological considerations. *International Journal of Performance Analysis in Sport, 12*(2), 398–424.

Bloomfield, J., Polman, R., & O'Donoghue, P. (2007). Physical demands of different positions in FA Premier League soccer. *Journal of Sports Science & Medicine, 6*(1), 233–242.

Bradley, P. S., Sheldon, W., Wooster, B., Olsen, P., Boanas, P., & Krustrup, P. (2009). High-intensity running in English FA Premier League soccer matches. *Journal of Sports Sciences, 27*(2), 159–168.

Buchheit, M., Allen, A., Poon, T. K., Modonutti, M., Gregson, W., & Di Salvo, V. (2014). Integrating different tracking systems in football: Multiple camera semi-automatic system, local position measurement and GPS technologies. *Journal of Sports Sciences, 00*(00), 1–14.

Buchheit, M., Mendez-villanueva, A., Simpson, B. M., & Bourdon, P. C. (2010). Repeated-sprint sequences during youth soccer matches. *International Journal of Sports Medicine, 31*(10), 709–716.

Carling, C. (2010). Analysis of physical activity profiles when running with the ball in a professional soccer team. *Journal of Sports Sciences, 28*(3), 319–326.

Carling, C. (2013). Interpreting physical performance in professional soccer match-play: Should we be more pragmatic in our approach? *Sports Medicine (Auckland, N.Z.), 43*(8), 655–663.

Carling, C., Bloomfield, J., Nelsen, L., & Reilly, T. (2008). The role of motion analysis in elite soccer. *Sports Medicine, 38*(10), 839–862.

Carling, C., Williams, A. M., & Reilly, T. (2005). *Handbook of soccer match analysis: A systematic approach to improving performance.* London & New York: Taylor & Francis Group.

Casamichana, D., Castellano, J., & Castagna, C. (2012). Comparing the physical demands of friendly matches and small-sided games in semiprofessional soccer players. *Journal of Strength and Conditioning Research, 26*(3), 837–843.

Chamari, K., Hachana, Y., Ahmed, Y. B., Galy, O., Sghaïer, F., Chatard, J.-C., ... Wisløff, U. (2004). Field and laboratory testing in young elite soccer players. *British Journal of Sports Medicine, 38*(2), 191–196.

Clemente, F. M., Couceiro, M. S., Fernando, M. L., Mendes, R., & Figueiredo, A. J. (2013a). Measuring tactical behaviour using technological metrics: Case study of a football game. *International Journal of Sports Science & Coaching, 8*(4), 723–739.

Clemente, F. M., Couceiro, M. S., Martins, F. M. L., Ivanova, M. O., & Mendes, R. (2013b). Activity profiles of soccer players during the 2010 world cup. *Journal of Human Kinetics, 38,* 201–211.

Clemente, F. M., Couceiro, M. S., Martins, F. M. L., Mendes, R. S., & Figueiredo, A. J. (2014a). Practical implementation of computational tactical metrics for the football game: Towards an augmenting perception of coaches and sport analysts. In Murgante, Misra, Rocha, Torre, Falcão, Taniar, ... Gervasi (Eds.), *Computational science and its applications* (pp. 712–727). Springer.

Clemente, F. M., Martins, F. M. L., Couceiro, M. S., Mendes, R., & Figueiredo, A. J. (2014b). Inspecting teammates' coverage during attacking plays in a football game: A case study Inspecting teammates' coverage during attacking plays in a football game : A case study. *International Journal of Performance Analysis in Sport, 14*(2), 1–27.

Clemente, F. M., Couceiro, M. S., Martins, F. M. L., Mendes, R. S., & Figueiredo, A. J. (2015a). Soccer team's tactical behaviour: Measuring territorial domain. *Proceedings of the Institution of Mechanical Engineers, Part P: Journal of Sports Engineering and Technology, 229*(1), 58–66.

Clemente, F. M., Martins, F. M. L., Kalamaras, D., Wong, D. P., & Mendes, R. S. (2015b). General network analysis of national soccer teams in FIFA World Cup 2014. *International Journal of Performance Analysis in Sport, 15*(1), 80–96.

Clemente, F. M., Martins, F. M. L., Wong, D. P., Kalamaras, D., & Mendes, R. S. (2015c). Midfielder as the prominent participant in the building attack: A network analysis of national teams in FIFA World Cup 2014. *International Journal of Performance Analysis in Sport, 15* (2), 704–722.

Datson, N., Hulton, A., Andersson, H., Lewis, T., Weston, M., Drust, B., & Gregson, W. (2014). Applied physiology of female soccer: An update. *Sports Medicine (Auckland, N.Z.), 44*(9), 1225–1240.

Dellal, A., Chamari, K., Pintus, A., Girard, O., Cotte, T., & Keller, D. (2008). Heart rate responses during small-sided games and short intermittent running training in elite soccer players: A comparative study. *Journal of Strength and Conditioning Research, 22*(5), 1449–1457.

Di Salvo, V., Baron, R., González-Haro, C., Gormasz, C., Pigozzi, F., & Bachl, N. (2010). Sprinting analysis of elite soccer players during European Champions League and UEFA Cup matches. *Journal of Sports Sciences, 28*(14), 1489–1494.

Di Salvo, V., Baron, R., Tschan, H., Calderon Montero, F. J., Bachl, N., & Pigozzi, F. (2007). Performance characteristics according to playing position in elite soccer. *International Journal of Sports Medicine, 28,* 222–227.

Duch, J., Waitzman, J. S., & Amaral, L. A. (2010). Quantifying the performance of individual players in a team activity. *PLoS ONE, 5*(6), e10937.

Ekblom, B. (1986). Applied physiology of soccer. *Sports Medicine, 3*(1), 50–60.

Frencken, W., Lemmink, K., Delleman, N., & Visscher, C. (2011). Oscillations of centroid position and surface area of football teams in small-sided games. *European Journal of Sport Science, 11*(4), 215–223.

Gerisch, G., Rutemöller, E., & Weber, K. (1988). Sports medical measurements of performance in soccer. In T. Reilly, A. Lees, K. Davids, & Y. W. Murphy (Eds.), *Science and football* (pp. 60–67). London, UK: E & FN Spon.

Gil, S., Gil, J., Ruiz, F., Irazusta, A., & Irazusta, J. (2007a). Physiological and anthropometric characteristics of young soccer players according to their playing position: Relevance for the selection process. *Journal of Strength & Conditioning Research, 21*(2), 438–445.

Gil, S. M., Gil, J., Ruiz, F., Irazusta, A., & Irazusta, J. (2007b). Physiological and anthropometric characteristics of young soccer players according to their playing position: Relevance for the selection process. *The Journal of Strength & Conditioning Research, 21*(2), 438–445.

Grund, T. U. (2012). Network structure and team performance: The case of English Premier League soccer teams. *Social Networks, 34*(4), 682–690.

Harley, J. A., Barnes, C. A., Portas, M., Lovell, R., Barrett, S., Paul, D., & Weston, M. (2010). Motion analysis of match-play in elite U12 to U16 age-group soccer players. *Journal of Sports Sciences, 28*(13), 1391–1397.

Haugen, T. A., Tønnessen, E., Hem, E., Leirstein, S., & Seiler, S. (2014). VO$_{2max}$ characteristics of elite female soccer players, 1989–2007. *International Journal of Sports Physiology and Performance, 9*(3), 515–521. doi:10.1123/IJSPP.2012-0150.

Hill-Haas, S. V., Coutts, A. J., Rowsell, G. J., & Dawson, B. T. (2009). Generic versus small-sided game training in soccer. *International Journal of Sports Medicine, 30*(9), 636–642.

Hoff, J., Wisløff, U., Engen, L. C., Kemi, O. J., & Helgerud, J. (2002). Soccer specific aerobic endurance training. *British Journal of Sports Medicine, 36.* doi:10.1136/bjsm.36.3.218

Hughes, M., & Franks, I. (2005). Analysis of passing sequences, shots and goals in soccer. *Journal of Sports Sciences, 23*(5), 509–514.

Jensen, K., & Larsson, B. (1993). Variation in physical capacity in a period including supplemental training of the national Danish soccer team for women. In T. Reilly, J. Clarys, & A. Stibbe (Eds.), *Science and football II* (pp. 114–117). London, UK: E&FN Spon.

Krustrup, P., Hellsten, Y., & Bangsbo, J. (2004). Intense interval training enhances human skeletal muscle oxygen uptake in the initial phase of dynamic exercise at high but not at low intensities. *The Journal of Physiology, 559*(1), 335–345.

Krustrup, P., Mohr, M., Ellingsgaard, H., & Bangsbo, J. (2005). Physical demands during an elite female soccer game: Importance of training status. *Medicine and Science in Sports and Exercise, 37*(7), 1242–1248.

Krustrup, P., Mohr, M., Steensberg, A., Bencke, J., Kjaer, M., & Bangsbo, J. (2006). Muscle and blood metabolites during a soccer game. *Medicine and Science in Sports and Exercise, 38*(6), 1165–1174.

Lago-Ballesteros, J., & Lago-Peñas, C. (2010). Performance in team sports: Identifying the keys to success in soccer. *Journal of Human Kinetics, 25,* 85–91.

Lago-Peñas, C., & Dellal, A. (2010). Ball possession strategies in elite soccer according to the evolution of the match-score: The influence of situational variables. *Journal of Human Kinetics, 25,* 93–100.

Lago-Peñas, C., & Lago-Ballesteros, J. (2011). Game location and team quality effects on performance profiles in professional soccer. *Journal of Sports Science and Medicine, 10,* 465–471.

Little, T. (2009). Optimizing the use of soccer drills for physiological development. *Strength and Conditioning Journal, 31*(3), 67–74.

Mendez-Villanueva, A., Buchheit, M., Simpson, B., & Bourdon, P. (2012). Match play intensity distribution in youth soccer. *International Journal of Sports Medicine, 34*(02), 101–110.

Mohr, M., Krustrup, P., Andersson, H., Kirkendal, D., & Bangsbo, J. (2008). Match activities of elite women soccer players at different performance levels. *Journal of Strength and Conditioning Research, 22*(2), 341–349.

Mohr, M., Krustrup, P., & Bangsbo, J. (2003). Match performance of high-standard soccer players with special reference to development of fatigue. *Journal of Sports Sciences, 21*(7), 519–528.

Mohr, M., Krustrup, P., & Bangsbo, J. (2005). Fatigue in soccer: a brief review. *Journal of Sports Sciences, 23*(6), 593–599.

Mohr, M., Krustrup, P., Nybo, L., Nielsen, J. J., & Bangsbo, J. (2004). Muscle temperature and sprint performance during soccer matches—beneficial effect of re-warm-up at half-time. *Scandinavian Journal of Medicine and Science in Sports, 14*(3), 156–162.

Moura, F. A., Martins, L. E., Anido, R. O., Barros, R. M., & Cunha, S. A. (2012). Quantitative analysis of Brazilian football players' organization on the pitch. *Sports Biomechanics, 11*(1), 85–96.

Ogushi, T., Ohashi, J., Nagahama, H., Isokawa, M., & Suzuki, S. (1993). Work intensity during soccer match-play (a case study). In T. Reilly, J. Clarys, & A. Stibbe (Eds.), *Science and football II* (pp. 121–123). London, UK: E&FN Spon.

Peña, J. L., & Touchette, H. (2012). A network theory analysis of football strategies. In *arXiv preprint arXiv* (p. 1206.6904).

Pollard, R. (2006). Home advantage in soccer: Variations in its magnitude and a literature review of the inter-related factors associated with its existence. *Journal of Sport Behavior, 29*, 169–189.

Polman, R., Walsh, D., Bloomfield, J., & Nesti, M. (2004). Effective conditioning of female soccer players. *Journal of Sports Sciences, 22*(2), 191–203.

Rampinini, E., Coutts, A. J., Castagna, C., Sassi, R., & Impellizzeri, F. M. (2007). Variation in top level soccer match performance. *International Journal of Sports Medicine, 28*(12), 1018–1024.

Rampinini, E., Impellizzeri, F. M., Castagna, C., Coutts, A. J., & Wisløff, U. (2009). Technical performance during soccer matches of the Italian Serie A league: effect of fatigue and competitive level. *Journal of Science and Medicine in Sport/Sports Medicine Australia, 12*(1), 227–233.

Reilly, T. (2007a). *The science of training—soccer: A scientific approach to developing strength, speed and endurance*. Abingdon, Oxon: Routledge, Taylor & Francis Group.

Reilly, T. (2007b). *The science of training—soccer*. Oxon, UK: Routledge.

Reilly, T., & Thomas, V. (1976). A motion analysis of work-rate in different positional roles in professional football match-play. *Journal of Human Movement Studies, 2*, 87–97.

Reilly, T., & Williams, A. M. (2003). *Science and soccer* (Second.). London, UK: Routledge Taylor & Francis Group.

Rienzi, V., Drust, B., Reilly, T., Carter, J., & Martin, A. (2000). Investigation of anthropometric and work-rate profiles of elite South American international soccer players. *The Journal of Sports Medicine and Physical Fitness, 40*(2), 162–169.

Souglis, A. G., Papapanagiotou, A., Bogdanis, G. C., Travlos, A. K., Apostolidis, N. G., & Geladas, N. D. (2015). Comparison of inflammatory responses to a soccer match between elite male and female players. *Journal of Strength and Conditioning Research, 29*(5), 1227–1233.

Sporis, G., Jukic, I., Ostojic, S. M., & Milanovic, D. (2009). Fitness profiling in soccer: Physical and physiologic characteristics of elite players. *Journal of Strength and Conditioning Research, 23*(7), 1947–1953.

Stølen, T., Chamari, K., Castagna, C., & Wisløff, U. (2005). Physiology of soccer. *Sports Medicine, 35*(6), 501–536.

Stroyer, J., Hansen, L., & Klausen, K. (2004). Physiological profile and activity pattern of young soccer players during match play. *Medicine and Science in Sports and Exercise, 36*(1), 168–174.

Tenga, A., Holme, I., Ronglan, L. T., & Bahr, R. (2010). Effect of playing tactics on achieving score-box possessions in a random series of team possessions from Norwegian professional soccer matches. *Journal of Sports Sciences, 28*(3), 245–255.

Turner, A. N., & Stewart, P. F. (2014). Strength and conditioning for soccer players. *Strength and Conditioning Journal, 36*(4), 1–13.

Vigne, G., Gaudino, C., Rogowski, I., Alloatti, G., & Hautier, C. (2010). Activity profile in elite Italian soccer team. *International Journal of Sports Medicine, 31*(05), 304–310.

Vilar, L., Araújo, D., Davids, K., & Bar-Yam, Y. (2013). Science of winning football: Emergent pattern-forming dynamics in association football. *Journal of Systems Science and Complexity, 26*, 73–84.

Wong, D. P., & Wong, S. H. (2009). Physiological profile of Asian elite youth soccer players. *Journal of Strength and Conditioning Research, 23*(5), 1383–1390.

Chapter 3
Small-Sided and Conditioned Games Versus Traditional Training Methods: A Review

Abstract Small-sided and conditioned games lead to different physiological stimuli and medium- to long-term conditioning effects. These effects are the main factors that bring some skepticism of using these games instead of traditional training methods. Therefore, the aim of this review was to constitute a useful synthesis of all the researches that compared conditioning effects of SSCGs with traditional training programs. The reviewed studies revealed that SSCGs and traditional running-based programs have similar effects on the improvement of the aerobic system, the anaerobic system, speed, and power.

Keywords Small-sided and conditioned games · Running-based training methods · Acute effects · Adaptations · Soccer · Football · Sports training

3.1 Introduction

The effects of small-sided and conditioned games (SSCGs) on exercise intensity experienced by players have been well researched in the recent years (Clemente et al. 2014a; Gabbett et al. 2009; Halouani et al. 2014b; Hill-Haas et al. 2011). The characterization of physiological, physical, technical, and tactical responses demonstrates their relevant contribution for training development in team sports (Aguiar et al. 2012; Hoffmann et al. 2014). In spite of that, the lack of consensus about the long-term physiological and physical effects of these games in comparison with traditionally run training programs is the reason that skepticism in the training context still remains. For that reason, the aim of this review is threefold: (i) compare the physiological and physical adaptations of SSCGs and high-intensity interval training programs; (ii) compare physiological and physical adaptations of other traditionally isolated training methods with SSCG programs; and (iii) identify the main physiological and physical adaptations to SSCG training programs. Hopefully, this review can contribute to reduce skepticism about the long-term benefits of SSCG training programs and to improve best practices regarding the conditioning of team sports players.

© The Author(s) 2016   27
F.M. Clemente, *Small-Sided and Conditioned Games in Soccer Training*,
SpringerBriefs in Applied Sciences and Technology,
DOI 10.1007/978-981-10-0880-1_3

The databases of MEDLINE/PubMed, SPORTDiscus, and Google Scholar were used to search for the literature. For scientific studies, only peer-reviewed articles in English were included. The following keywords were used in multiple combinations: "small-sided games," "small-sided and conditioned games," "physiological and physical effects," "teams-specific conditioning," "high-intensity interval training," "skill-based training," "skill-based conditioning," and "traditional running methodologies." Due to the focus on team sports and the lack of studies on long-term effects, no limit to the search period was applied.

3.2 Variables Affecting Small-Sided Games' Intensity

SSCGs are smaller and adapted versions of games that are often used in the context of training for team sports as a part of their regular programs (Clemente et al. 2014b). SSCGs optimize the time of training session, allowing for simultaneous development of physiological, physical, and technical/tactical performance in the same exercise (Dellal et al. 2012). The advantage of SSCGs has been examined with respect to their potential to improve aerobic fitness (Delextrat and Martínez 2014; Hill-haas et al. 2009b) and to attain an exercise intensity of 85–90 % of HRmax, blood lactate concentrations of 4–8 mmol/L, and a rating of perceived exertion (RPE) of 6–8 (on a scale of 10). Moreover, some studies also showed that some SSCG formats resulted in HR responses comparable with short-duration intermittent running (Halouani et al. 2014b). These physiological and physical responses are constrained by the variables used by coaches during the design of SSCGs (Davids et al. 2013). For that reason, the following will describe the general effects of different variables and conditions on physiological and physical responses during SSCGs.

3.2.1 Formats

Generally, the studies conducted in team sports (particularly in soccer, basketball, and rugby) have revealed that SSCGs with a smaller number of players (smaller formats) statistically increase the heart rate responses, blood lactate concentrations, and perceived exertion than games with greater number of players (Hill-haas et al. 2009a; Kennett et al. 2012; Klusemann et al. 2012; Köklü et al. 2011; Owen et al. 2011; Rampinini et al. 2007). Such evidence is common to different team sports and the range of values found in the most relevant studies can be found in Table 3.1.

Briefly, smaller formats (one-on-one and two-on-two) can promote values of approximately 90 % HRmax and blood lactate concentrations of about 8 mmol/L. The range of values for these smaller formats is quite large and may be constrained by the regimen of training (continuous vs. intermittent) and the size of the playing area

Table 3.1 Summary of studies examining the effects of format on SSCGs' intensity in team sports

Studies	Team sport	Format	Range of % HRmax (%)	Range of BLa⁻ (mmol/L)	Range of RPE (scale of 10)
Aroso et al. (2004), Brandes et al. (2012), Dellal et al. (2011), Hill-Haas et al. (2009a), Köklü (2012), Rampinini et al. (2007)	Soccer	1 versus 1	86.9–89.0	9.4–9.4	–
		2 versus 2	80.1–93.3	3.5–8.1	7.6
		3 versus 3	81.7–94.0	3.3–7.5	7.7
		4 versus 4	70.6–91.5	2.6–6.9	7.9
		5 versus 5	75.7–92.7	2.5–5.2	13.48 (0–20)
		6 versus 6	82.8–88.0	2.6–5.0	–
Castagna et al. (2011), Conte et al. (2015a), Delextrat and Kraiem (2013), Klusemann et al. (2012), McCormick et al. (2012), Sampaio et al. (2009)	Basketball	1 versus 1	–	–	–
		2 versus 2	86.0–89.9	7.8	6.8–8.8
		3 versus 3	87.1–88.0	6.2	3.0–5.8
		4 versus 4	82.7–87.3	–	4.1–7.7
Foster et al. (2010), Kennett et al. (2012)	Rugby	4 versus 4	87.9–91.5	8.9	17.4 (0–20)
		6 versus 6	88.4–90.3	6.5	15.0 (0–20)

(smaller or larger) (Halouani et al. 2014b; Hill-Haas et al. 2011). These formats are more adequate for the development of glycolic system/anaerobic training and also to increase the intensity of running and the ballistic movements (sprints, high-intensity shuffling movements, or jumps) (Clemente et al. 2014a; Klusemann et al. 2012).

On the other hand, bigger SSCGs (five-on-five and six-on-six in soccer and rugby; three-on-three and four-on-four in basketball) induce values of approximately 85 % of HRmax and blood lactate concentrations of 4.5 mmol/L, thus suggesting an activity profile more adequate to developing a high-intensity aerobic system with a continuous regimen or an intermittent regimen with longer periods of practice (Little 2009; Sassi et al. 2004). These games are also associated with a decrease in the intensity/velocity of running (Delextrat and Kraiem 2013).

3.2.2   Field and Court Dimensions

The majority of studies that compared different dimensions of the field revealed that having larger dimensions increases the HR responses, blood lactate concentrations, and perceived exertion (Atli et al. 2013; Casamichana and Castellano 2010; Kennett et al. 2012; Klusemann et al. 2012; Owen et al. 2004; Williams and Owen 2007). Table 3.2 summarizes the studies that have examined the influence of field and court area on SSCGs' intensity in three team sports (soccer, basketball, and rugby).

Based on the evidence in different team sports, it is possible to assume that larger field dimensions increase the physiological responses because of the increased space each player must cover and the decreased opportunity for recovery (Clemente et al. 2014a). In the case of studies that analyzed the performance responses between different field sizes, the influence of different formats was also investigated. For that reason, it is possible to identify a concurrent manipulation of both variables (Hill-Haas et al. 2011). The common evidence reveals that decreasing player numbers with a constant field area per player and smaller player numbers with a large field area are both suitable methods for increasing the intensity of SSCGs (Halouani et al. 2014b).

Table 3.2 Summary of studies examining the effects of field and court dimensions on SSCGs' intensity in team sports

Studies	Team sport	Format	Field area	Range of % HRmax (%)	Range of BLa⁻ (mmol/L)	Range of RPE (scale of 10)
Aroso et al. (2004), Casamichana and Castellano (2010), Kelly and Drust (2009), Owen et al. (2011), Rampinini et al. (2007), Williams and Owen (2007)	Soccer	1 versus 1 to 2 versus 2	Smaller	84.2–88.0	–	–
			Medium	87.4–89.0	–	–
			Larger	88.1–89.0	–	–
		3 versus 3 to 4 versus 4	Smaller	72.0–89.5	2.6–6.0	7.6–8.1
			Medium	78.5–90.5	5.5–6.3	7.2–8.4
			Larger	75.7–94.0	6.0–6.5	8.1–8.5
		5 versus 5 to 6 versus 6	Smaller	79.5–93.0	4.5–5.0	5.7–7.3
			Medium	86.4–94.6	5.0–5.0	6.7–7.6
			Larger	80.2–94.6	4.8–5.8	6.7–7.5
Atli et al. (2013), Klusemann et al. (2012)	Basketball	1 versus 1 to 2 versus 2	Smaller	84	–	8.3
			Larger	85	–	8.6
		3 versus 3 to 4 versus 4	Smaller	76.3	–	6.3
			Larger	85.6	–	7.7
Foster et al. (2010), Kennett et al. (2012)	Rugby	3 versus 3 to 4 versus 4	Smaller	87.9–89.8	–	–
			Larger	88.4–91.5	5.7	13.7 (0–20)
		5 versus 5 to 6 versus 6	Smaller	88.5–88.5	–	–
			Larger	89.4–90.3	8.2	15.8 (0–20)

3.2.3   Rule Modifications and Tactical Guidelines

The modification of rules and specific tactical guidelines is commonly used by coaches to augment the perception of players for a given technical action or tactical content (Davids et al. 2013). Similar to format and size of the field, the changes in rules also lead to different physiological, physical, and technical/tactical performance (Clemente et al. 2014a). Table 3.3 shows the effects of different types of rule

Table 3.3 Summary of studies examining the effects of rule modifications and tactical guidelines on SSCGs' intensity in team sports

Studies	Team sport	Format	Rule modifications	% HRmax	BLa⁻ (mmol/L)	RPE (scale of 10)
Aroso et al. (2004)	Soccer	2 versus 2	Zone Marking	84.0 %	8.1	16.2 (0–20)
		2 versus 2	Man-to-man marking	82.9 %	9.7	16.7 (0–20)
		3 versus 3	Free Touch	86.8 %	4.9	14.5 (0–20)
		3 versus 3	Maximum of 3 touches on ball	85.3 %	5.3	15.4 (0–20)
Sassi et al. (2004)		4 versus 4	With GK	174 HRmean	–	–
			Without GK	178 HRmean	–	–
		8 versus 8 with GK	Free Touch	82.0 %	–	–
		8 versus 8 with GK	Free touch with pressure	91.0 %	–	–
Little and Williams (2006)		5 versus 5	Pressure half switch	89.9 %	–	–
		6 versus 6	Pressure half switch	90.5 %	–	–
Mallo and Navarro (2008)		3 versus 3	Possession	91.0 %	–	–
			Possession with outside players	91.0 %	–	–
			Regular with GK	88.0 %	–	–
Dellal et al. (2011)[a]		2 versus 2	1 touch	90.3 %	3.9	8.3
			2 touch	90.1 %	3.5	7.8
			Free touch	90.0 %	3.5	7.7
		3 versus 3	1 touch	90.0 %	3.6	8.2
			2 touch	89.4 %	3.4	7.9
			Free touch	89.6 %	3.1	7.5
		4 versus 4	1 touch	87.6 %	3.0	8.0
			2 touch	85.6 %	2.9	7.9
			Free touch	84.7 %	2.8	7.3

(continued)

Table 3.3 (continued)

Studies	Team sport	Format	Rule modifications	% HRmax	BLa⁻ (mmol/L)	RPE (scale of 10)
Jake et al. (2012)		3 versus 3	Man marking with goals	80.5 %	–	7.1
			Man Marking without goals	80.5 %	–	7.4
			Goals without man marking	75.7 %	–	6.9
			Without goals and without man marking	76.1 %	–	6.9
Castellano et al. (2013)		3 versus 3	SSG-P	94.6 %	–	–
			SSG-G	94.8 %	–	–
			SSG-g	91.8 %	–	–
		5 versus 5	SSG-P	94.6 %	–	–
			SSG-G	92.1 %	–	–
			SSG-g	91.5 %	–	–
		7 versus 7	SSG-P	94.9 %	–	–
			SSG-G	93.2 %	–	–
			SSG-g	94.7 %	–	–
Clemente et al. (2014a)		2 versus 2	Cross endline	74.98 % HRres	–	–
			Cross two-goals	81.05 % HRres	–	–
			Cross one-goal	83.38 % HRres	–	–
		3 versus 3	Cross endline	82.06 % HRres	–	–
			Cross two-goals	84.18 % HRres	–	–
			Cross one-goal	81.98 % HRres	–	–
		4 versus 4	Cross endline	81.27 % HRres	–	–
			Cross two-goals	80.32 % HRres	–	–
			Cross one-goal	83.61 % HRres	–	–
Conte et al. (2015b)	Basketball	4 versus 4	No dribble game	92.0 %	–	8.5
			Regular game	90.0 %	–	7.9

[a]The values only represent the professional players

changes such as the number of ball touches, the presence of a goalkeeper, the type of defense, or the type of score on physiological responses.

One of the main rule changes tested in recent years is the limitation of ball touches (Aroso et al. 2004; Dellal et al. 2011; Sassi et al. 2004). This condition aims to increase the speed of play and ball circulation. Moreover, it is also used to avoid players' individualism (Clemente et al. 2014a). In the majority of studies carried out in soccer, it is possible to observe that smaller numbers of touches gradually increase the heart rate responses and significantly increase the blood lactate concentration. In a similar study adapted to basketball, it was found that drills with no dribbling increased the heart rate responses in comparison with drills that did not limit dribbling (Conte et al. 2015b).

Another typical condition tested in the SSCGs studies is the use of goalkeepers (GKs) on soccer drills (Jake et al. 2012; Mallo and Navarro 2008; Sassi et al. 2004). The findings are relatively consistent in suggesting that the use of GKs decreases the heart rate responses and blood lactate concentration. The reason for this may be the tendency to increase the defensive organization to better protect the goal during games with goalkeepers (Mallo and Navarro 2008). Without using GKs, a study compared the use of different methods of scoring (Clemente et al. 2014a). Results showed that in two-on-two and 4 versus 4 cross one-goal (with possession of the ball and in dribble) it was more intense than using all endline to score.

The use of tactical guidelines to constrain players' behavior is also typically used in team sports. In soccer and basketball, some studies have analyzed the use of specific defensive conditions to influence the physiological and physical responses (Abdelkrim et al. 2010; Aroso et al. 2004; Jake et al. 2012; Sampaio et al. 2015). In the case of basketball, it compared man-to-man and zone defense (Abdelkrim et al. 2010) and found no statistical differences in the effects on heart rate responses and blood lactate concentration. In the case of soccer, a similar approached identified a gradual increase in heart rate responses for man-to-man defense (Jake et al. 2012). In an older study, it found the opposite in a two-on-two format (Aroso et al. 2004). Therefore, in basketball and soccer changes influenced by the type of defense played are not clear.

3.2.4 Coach Encouragement

The encouragement provided by coach supervision has also been tested in some studies on SSCGs in soccer and basketball (Gracia et al. 2014; Rampinini et al. 2007; Sampaio et al. 2007). In a study carried out with 20 amateur soccer players, significantly higher intensities of heart rate, blood lactate concentrations, and perceived exertion on the games were found with directed supervision and encouragement provided by the coach (Rampinini et al. 2007). Specifically, the average of HR responses (88.7 vs. 86.5 %), blood lactate concentration (5.5 vs. 4.2 mmol/L), and RPE (7.7 vs. 6.3) were significantly higher in the presence of encouragement (Rampinini et al. 2007). Similar evidence was found in two-on-two (83.7 vs.

81.2 % HRmax; 15.5 vs. 14.1 RPE) and three-on-three soccer formats (80.8 vs. 79.8 % HRmax; 15.8 vs. 14.4 RPE) (Sampaio et al. 2007).

In the case of study conducted in basketball (Gracia et al. 2014), it was found that in a three-on-three format with encouragement there was 68 % (in players under age 14, or U14) and 56.9 % (U16) of time in vigorous intensity (>85 % HRmax) and without encouragement the values decreased to 40.8 % (U14) and 47.7 % (U16). Conversely, in the same study it was found that in a four-on-four format without encouragement there were higher values of vigorous intensity (52.5 % in U14 and 74.0 % in U16) than with encouragement (47.0 % in U14 and 67.7 % in U16). In spite of this last evidence, it seems that encouragement provided by coaches during SSCGs can be a variable that increases the physiological intensity in team sports.

3.2.5 Intermittent and Continuous Regimens

The work-to-rest ratio can be a determinant variable that influences the physiological responses during SSCGs (Casamichana et al. 2013; Hill-haas et al. 2009a; Köklü 2012). The prescription of these games requires the following information (Halouani et al. 2014b): (i) work intensity and duration; (ii) recovery type (passive/active) and duration; and (iii) total work duration (work interval number × work duration). The majority of studies on SSCGs has used the intermittent regimen. Nevertheless, it is important to identify the different effects induced by continuous and intermittent regimens. Based on the studies that compared both regimens, Table 3.4 summarizes the main evidence about this issue in soccer and basketball.

In study that compared intermittent versus continuous training regimens in U16 soccer players playing two-on-two and four-on-four formats, no statistical differences were found between regimens for the distance covered while walking, jogging, or running at moderate speed (Hill-haas et al. 2009a). Nevertheless, the same study showed that players covered a significantly greater distance at 13.0–17.9 km/h, spent more time in velocities greater than 18 km/h, and performed a greater number of sprints in intermittent regimens. Conversely, significantly higher percentages of HRmax and RPE were verified in continuous regimens (Hill-haas et al. 2009a). A similar study carried out in U17 using three formats (teams of two, three, and four) revealed no statistical differences between regimens in percentages of HRmax and only in the case of two-on-two was a significant increase of blood lactate concentration found in continuous regimens. Finally, a study conducted in male soccer players (approximately 21 years old) revealed no significant differences in physiological and time–motion responses between two types of intermittent SSCGs and a continuous regimen (Casamichana et al. 2013).

The studies that compared short and long intermittent regimens (Conte et al. 2015a; Fanchini et al. 2011; Klusemann et al. 2012) in basketball and soccer revealed that longer intermittent regimens statistically increased the percentages of

Table 3.4 Summary of studies examining the effects of training regimens on SSCGs' intensity in team sports

Study	Team sport	Format	Prescription	Regimen	% HRmax
Hill-haas et al. (2009a)*	Soccer	2 versus 2	4 × 6 min/90-s rest	Interval	Interval 84.0 ± 1.0 % HRmax Continuous 87.0 ± 1.0 % HRmax
			1 × 24 min	Continuous	
		4 versus 4	4 × 6 min/90-s rest	Interval	
			1 × 24 min	Continuous	
		6 versus 6	4 × 6 min/90-s rest	Interval	
			1 × 24 min	Continuous	
Köklü (2012)		2 versus 2	3 × 2 min/2 min rest	Interval	88.6 ± 3.2
			1 × 6 min	Continuous	88.8 ± 3.2
		3 versus 3	3 × 3 min/2 min rest	Interval	92.0 ± 2.0
			1 × 9 min	Continuous	91.2 ± 2.6
		4 versus 4	3 × 4 min/2 min rest	Interval	90.1 ± 2.5
			1 × 12 min	Continuous	89.3 ± 2.7
Fanchini et al. (2011)**		3 versus 3	3 × 2 min/2 min rest	Interval	All exercise: 82.4 ± 4.1 Excluding the first minute: 88.5 ± 3.2
			3 × 4 min/2 min rest	Interval	All exercise: 85.9 ± 4.1 Excluding the first minute: 89.5 ± 3.1
			3 × 6 min/2 min rest	Interval	All exercise: 85.6 ± 3.9 Excluding the first minute: 87.8 ± 2.8
Casamichana et al. (2013)***		5 versus 5	4 × 4 min/1 min rest	Interval	87.5 ± 4.8
			2 × 8 min/2 min rest	Interval	87.1 ± 3.1
			1 × 16 min/2 min rest	Continuous	87.5 ± 3.1
Klusemann et al. (2012)*	Basketball	2 versus 2 and 4 versus 4	4 × 2.5 min/1 min rest	Short interval	83.0 ± 3.0
			2 × 5 min/30-s rest	Long interval	86.0 ± 4.0
Conte et al. (2015a)*		2 versus 2 and 4 versus 4	3 × 7 min/1 min exercise and 1 min rest	Long interval	86.5 ± 3.7
			3 × 4 min/2 min rest	Short interval	90.8 ± 2.7

*The available results do not differ the formats
**Results from the pooled version of three bouts
***Results from the pooled version of four periods of time

HRmax and RPE while shorter regimens significantly increased the skills carried out in a game or match.

For these reasons, there is no solid evidence that identifies differences between intermittent and continuous regimens. Nevertheless, the minor evidence suggests that intermittent regimens decrease the heart rate responses and blood lactate concentration (particularly the latter) and allow players to avoid fatigue for a long time.

3.3 Physiological and Physical Development After SSCG Training

The majority of studies in the field of SSCGs has analyzed the acute physiological and physical responses (Clemente et al. 2014a; Halouani et al. 2014b; Hill-Haas et al. 2011). Nevertheless, the effects after training programs seem to be the most important analysis to identify the value of this approach for team sports training. Table 3.5 summarizes the conditioning effects of SSCG training programs in team sports. The studies that compared SSCGs and other training methods were not included in this table.

The three studies conducted in different team sports (soccer, futsal, and rugby) revealed the tendency of SSCGs training programs to improve players' aerobic systems and their ability to repeat sprints (Berdejo-del-Fresno et al. 2015; Owen et al. 2012; Seitz et al. 2014). Nevertheless, these studies do not compare the SSCGs' effects with other training methods. For that reason, the majority of studies dedicated to conditioning effects have compared traditional training methods (Buchheit et al. 2009b; Delextrat and Martínez 2014; Sassi et al. 2004). Therefore, the following two sections of this review will analyze the acute responses and conditioning effects of SSCGs in comparison with high-intensity interval training and mixed generic training methods.

3.4 Comparing SSCGs with High-Intensity Interval Training (HIIT)

Despite the characterization of SSCGs' effects on acute physiological responses, one of the main contributions of these games is the possibility of them being used in training periodization to develop the conditioning of players. In fact, several researchers have questioned effectiveness of SSCGs when compared to traditionally run training programs of conditioning (Halouani et al. 2014a; Impellizzeri et al. 2005). To reduce skepticism, in recent years some studies have been conducted to compare the acute responses and training effects of SSCG programs with high-intensity interval training (HIIT) (Delextrat and Martínez 2014; Dellal et al.

Table 3.5 Summary of studies that compared pre- and post-SSCG training programs with conditioning variables

Study	Team sport	Sample	Training intervention	Preintervention	Postintervention
Owen et al. (2012)	Soccer	15 (elite)	4-weeks in season 7 SSG training intervention sessions Each lasting 45–90 minutes	Repeated Sprint Ability 10-m sprint 1.77 ± 0.07 s 20-m sprint 3.08 ± 0.11 s VO_{2max} (14 km/h) 52.03 ± 3.88 mL/kg/min HR (14 km/h) 184.43 ± 9.46 bpm	Repeated Sprint Ability 10-m sprint $1.75 \pm .05$ s 20-m sprint 3.06 ± 0.09 s VO_{2max} (14 km/h) 49.81 ± 4.54 mL/kg/min HR (14 km/h) 173.36 ± 11.38 bpm
Berdejo-del-Fresno et al. (2015)	Futsal (indoor soccer)	24 (elite)	6-weeks in season One session per week All the sessions consisted of 3 blocks of workout. Each block had a duration of 8 min, including the exercise time and the recovery time. The first week started with 4 min of workout and 4 min of rest; in every next week until the 4th week the workout time increased 1 min and the recovery time decreased 1 min	Agility (s) 9.49 ± 0.34 s Agility Ball (s) 11.29 ± 0.49 s Bleep Test (stages) 12.71 ± 0.80 VO_{2max} 58.73 ± 2.41 mL/kg/min	Agility (s) 9.38 ± 0.38 s Agility Ball (s) 11.21 ± 0.38 s Bleep Test (stages) 13.17 ± 1.00 VO_{2max} 60.11 ± 2.99 mL/kg/min
Seitz et al. (2014)	Rugby	10 (20.9 years old)	8-weeks Two sessions per week 16 SSCGs were played 7 different SSCGs were used 4×10 min/3 min recovery	Intermittent Shuttle Running V_{IFT} 19.35 ± 1.00 km/h 10-m sprint 1.95 ± 0.07 s 20-m sprint 3.28 ± 0.10 s 40-m sprint 5.34 ± 0.16 s	Intermittent Shuttle Running V_{IFT} 19.60 ± 0.77 km/h 10-m sprint 1.89 ± 0.06 s 20-m sprint 3.24 ± 0.08 s 40-m sprint 5.28 ± 0.13 s

2008; Impellizzeri et al. 2006b). Table 3.6 summarizes the studies that compared the acute physiological responses to both types of training.

A study of elite soccer players (Dellal et al. 2008) revealed that HR responses were significantly greater during 30–30-s AR test than during one-on-one, four-on-four, eight-on-eight, and ten-on-ten SSCGs and 10–10-s also had significantly greater HR responses than one-on-one, eight-on-eight, and ten-on-ten SSCGs. Another interesting result was the intersubject coefficient of variation for SSCGs (8.79–15.60 %) and HIIT (4.50–8.50 %). As expected, the HIIT method reduces the intervariability of responses for the controlled and standardized approach in comparison with SSCGs. On the other hand, the results of a comparative study with nine players showed that a four-on-four format had higher values of HR responses in comparison with runs of 4 × 1000 m with 150 s between bouts while higher blood lactate concentrations were found in the HIIT method (Sassi et al. 2004). Finally, in a study conducted in handball, significantly greater values of blood lactate concentrations and percentages of anaerobic energy in HIIT were found in comparison with four-on-four games (Buchheit et al. 2009b).

If the previous studies analyzed the acute physiological responses to SSCGs and HIIT training methods, it is more important to identify the training effects on players' conditioning. These medium- to long-term effects are the main priority for the periodization and validation of methodological approaches. For that reason, the SSCG and HIIT programs have been compared by their effects after intervention (Buchheit et al. 2009b; Delextrat and Martínez 2014; Jastrzebski et al. 2014). The summary of studies that analyzed the effects of both programs is presented in Table 3.7.

The conditioning effects of SSCGs and HIIT were analyzed in soccer, handball, and basketball. In the majority of studies, no significant statistical differences were found between methods in aerobic/anaerobic systems and physical capacities (Buchheit et al. 2009b; Delextrat and Martínez 2014; Impellizzeri et al. 2006b) in any of the three team sports.

In the case of pre- and postintervention (6–12 weeks) effects, it was found that both SSCGs and HIIT produce significant improvements in aerobic responses measured by VO_{2max} (7–8 % better) (Impellizzeri et al. 2006a; Radziminski et al. 2013; Reilly and White 2004), V_{O2} at the lactate threshold (8–13 % better) (Impellizzeri et al. 2006a; Radziminski et al. 2013), and 30-15$_{IFT}$ (3–6 % better) (Buchheit et al. 2009b; Delextrat and Martínez 2014). In the case of speed, both training programs revealed statistical improvements in the 5-m sprint (2–3 % faster) (Radziminski et al. 2013), 10-m sprint (1–4 % faster) (Buchheit et al. 2009b; Iacono et al. 2015), 20-m sprint (2–4 % faster) (Iacono et al. 2015), and agility (1–2 % faster) (Iacono et al. 2015). Finally, the effects of both training programs on countermovement jumps (3–10 % better) (Arcos et al. 2015; Buchheit et al. 2009b; Iacono et al. 2015), peak power (4–5 % better) (Jastrzebski et al. 2014; Radziminski et al. 2013), total work capacity (4–5 % better) (Radziminski et al. 2013), bench pressing (6–12 % better) (Iacono et al. 2015), upper body power (1–7 % better) (Delextrat and Martínez 2014), and lower body power (1–4 % better) (Delextrat and Martínez 2014) have also been analyzed.

Table 3.6 Summary of studies comparing the acute physiological responses of SSCGs and HIIT training methods in team sports

Study	Team sport	Sample	Group	Training intervention	Results
Sassi et al. (2004)	Soccer	9	HIIT	4 × 1000-m runs with 150 s between bouts	HRaverage 167 bpm
					Bla⁻ 7.9 ± 3.4 mmol/L
			SSCG	4 versus 4	With GK HRaverage 174 bpm
					Bla⁻ 6.2 ± 1.4 mmol/L
					Without GK HRaverage 178 bpm
					Bla⁻ 6.4 ± 2.7 mmol/L
				8 versus 8	HRaverage 160 bpm
					Bla⁻ 3.3 ± 1.2 mmol/L
Dellal et al. (2008)*		10 (26.0 ± 2.9 years old)	HIIT	10-10PR 2 × 7 min; Total duration 20 min; 6 min Interseries recovery	% HRres 85.8 ± 3.9
				30-30PR 2 × 10 min; Total duration 30 min; 10 min Interseries recovery	% HRres 77.2 ± 4.6
				30-30AR 2 × 10 min; Total duration 30 min; 10 min Interseries recovery	% HRres 85.7 ± 4.5
				15-15PR 2 × 10 min; Total duration 28 min; 8 min Interseries recovery	% HRres 76.8 ± 4.0
				5-20PR 1 × 7 min, 5 s; Total duration 7 min	% HRres 80.2 ± 6.8

(continued)

Table 3.6 (continued)

Study	Team sport	Sample	Group	Training intervention	Results
			SSCG	1 versus 1 4 × 1 min, 30 s; Total duration 10 min and 30 s; 1 min and 30 s Interseries recovery	% HRres 77.6 ± 8.6
				2 versus 2 6 × 2 min, 30 s; Total duration 27 min and 30 s; 2 min and 30 s Interseries recovery	% HRres 80.1 ± 8.7
				4 versus 4 with GK 2 × 4 min; Total duration 11 min; 3 min Interseries recovery	% HRres 77.1 ± 10.7
				8 versus 8 with GK 2 × 10 min; Total duration 25 min; 5 min Interseries recovery	% HRres 80.3 ± 12.5
				8 versus 8 4 × 4 min; Total duration 25 min; 3 min Interseries recovery	% HRres 71.7 ± 6.3
				10 versus 10 with GK 3 × 20 min; Total duration 70 min; 5 min Interseries recovery	% HRres 75.7 ± 7.9
Buchheit et al. (2009b)	Handball	9 (21 years old)	HIIT	Running 8-min in 15 s effort (95 % V_{IFT}) interspersed with 15 s of passive recovery	VO_{2peak} 56.4 mL/min/Kg HRpeak 189 bpm BLa^- 11.6 mmol/L
			4 versus 4	2 × 225 s/30 s rest	VO_{2peak} 60.2 mL/min/Kg HRpeak 187 bpm BLa^- 8.9 mmol/L

*PR Passive recovery (rest); AR Active recovery

Table 3.7 Summary of studies comparing the effects of SSCGs and HIT training programs in team sports

Study	Team sport	Sample	Group/period	Training intervention	Results
Reilly and White (2004)	Soccer	18 (professional U18)	HIT 6-weeks	2 days per week Running 6 × 4 min (85–90 % HRmax)/3 min active recovery	Preintervention VO$_{2max}$ 57.8 ± 3.2 mL/kg/min Postintervention VO$_{2max}$ 58.0 ± 3.2 mL/kg/min
			SSCG 6-weeks	2 days per week 5 versus 5 6 × 4 min/3 min active recovery	Preintervention VO$_{2max}$ 57.7 ± 3.0 mL/kg/min Postintervention VO$_{2max}$ 57.8 ± 3.2 mL/kg/min
Impellizzeri et al. (2006a)		40 (junior)	HIT 12-weeks	2 days per week Running 4 × 4 min (90-95 % HRmax)/3 min active recovery	Preintervention VO$_{2max}$ 55.6 ± 3.4 mL/kg/min V$_{O2}$ at Lactate Threshold 45.1 ± 3.4 mL/kg/min High Intensity in match 351 ± 67 s Postintervention VO$_{2max}$ 60.2 ± 3.9 mL/kg/min V$_{O2}$ at Lactate Threshold 50.9 ± 2.9 mL/kg/min High Intensity in match 431 ± 75 s
			SSCG 12-weeks	2 days per week 3 versus 3, 4 versus 4 and 5 versus 5 4 × 4 min/3 min active recovery	Preintervention VO$_{2max}$ 57.7 ± 4.2 mL/kg/min V$_{O2}$ at Lactate Threshold

(continued)

Table 3.7 (continued)

Study	Team sport	Sample	Group/period	Training intervention	Results
					47.3 ± 4.9 mL/kg/min
					High Intensity in match
					377 ± 60 s
					Postintervention
					VO_{2max}
					61.8 ± 4.5 mL/kg/min
					V_{O2} at Lactate Threshold
					52.4 ± 2.8 mL/kg/min
					High Intensity in match
					473 ± 89 s
Dellal et al. (2012)		22 (26.3 years old amateur)	HIT 6-weeks	2 days per week	Preintervention
				Running	HRres
				10 s–10 s	137 ± 5 b min^{-1}
				2 blocks 2 × 7 (95 % V30-15$_{IFT}$)/5 min interseries recovery and 7 × 6 recovery periods	ISCV/end-speed velocity at Vameval
				15 s–15 s	4.43
				2 blocks 2 × 8 (100 % V30-15$_{IFT}$)/5 min interseries recovery and 2 × 7 recovery periods	ISCV-V30-15$_{IFT}$
				30 s–30 s	3.09
				2 blocks 2 × 10 (95 % V30-15$_{IFT}$)/6 min interseries recovery and 2 × 9 recovery periods	Postintervention
					HRres
					139 ± 4 b min^{-1}
					ISCV/end-speed velocity at Vameval
					4.09
					ISCV-V30-15$_{IFT}$
					6.20
			SSCG 6-weeks	2 days per week	Preintervention
				1 versus 1	HRres
				5 × 1 min and 30 s/1.5 min recovery and	130 ± 6 b min^{-1}

(continued)

Table 3.7 (continued)

Study	Team sport	Sample	Group/period	Training intervention	Results
				2 versus 2	ISCV/end-speed velocity at Vameval
				5 × 2 min and 30 s/2 min recovery	3.49
					ISCV-V30-15$_{IFT}$
					4.81
					Postintervention
					HRres
					131 ± 5 b min^{-1}
					ISCV/end-speed velocity at Vameval
					5.63
					ISCV-V30-15$_{IFT}$
					6.20
Radziminski et al. (2013)		20 (U16)	HIT 8-weeks	2 days per week	Preintervention
				Running	VO$_{2max}$
				5 × 4 min (90 % HRmax)/3 min active recovery	56.2 ± 8.67 mL/kg/min
					V$_{O2}$ at Lactate Threshold
					44.7 ± 12.48 mL/kg/min
					Peak Power
					10.8 ± 1.02 W kg^{-1}
					Total Work Capacity
					258.5 ± 15.39 J kg^{-1}
					5-m sprint
					1.21 ± 0.05 s
					Postintervention
					VO$_{2max}$
					55.3 ± 6.07 mL/kg/min
					V$_{O2}$ at Lactate Threshold
					42.9 ± 9.23 mL/kg/min
					Peak Power
					11.3 ± 0.82 W kg^{-1}
					Total Work Capacity
					270.3 ± 15.11 J kg^{-1}
					5-m sprint
					1.17 ± 0.06 s

(continued)

Table 3.7 (continued)

Study	Team sport	Sample	Group/period	Training intervention	Results
			SSCG 8-weeks	2 days per week 3 versus 3 with and without neutral player 5 × 4 min/3 min active recovery	Preintervention VO$_{2max}$ 58.6 ± 6.93 mL/kg/min V$_{O2}$ at Lactate Threshold 46.17 ± 9.16 mL/kg/min Peak Power 10.6 ± 0.94 W kg^{-1} Total Work Capacity 261.3 ± 18.09 J kg^{-1} 5-m sprint 1.22 ± 0.06 s Postintervention VO$_{2max}$ 63.3 ± 8.04 mL/kg/min V$_{O2}$ at Lactate Threshold 49.70 ± 11.64 mL/kg/min Peak Power 11.2 ± 1.02 W kg^{-1} Total Work Capacity 272.8 ± 20.71 J kg^{-1} 5-m sprint 1.19 ± 0.06 s
Jastrzębski et al. (2014)		22 (U16)	HIT 8-weeks	2 days per week Running 7 × 3 min (85–90 % HRmax)/90 s active recovery	Preintervention VO$_{2max}$ 55.7 ± 5.23 mL/kg/min Peak Power 11.0 ± 0.98 W kg^{-1} 5-m sprint 1.16 ± 0.06 s 30-m sprint 4.66 ± 0.22 s

(continued)

Table 3.7 (continued)

Study	Team sport	Sample	Group/period	Training intervention	Results
					Postintervention
					VO_{2max}
					56.9 ± 5.58 mL/kg/min
					Peak Power
					11.3 ± 1.02 W kg^{-1}
					5-m sprint
					1.16 ± 0.08 s
					30-m sprint
					4.62 ± 0.22 s
			SSCG	2 days per week	Preintervention
			8-weeks	3 versus 3	VO_{2max}
				4×3 min/90 s active recovery	52.5 ± 5.15 mL/kg/min
					Peak Power
					10.7 ± 0.78 W kg^{-1}
					5-m sprint
					1.18 ± 0.05 s
					30-m sprint
					4.61 ± 0.25 s
					Postintervention
					VO_{2max}
					57.0 ± 5.44 mL/kg/min
					Peak Power
					11.0 ± 0.72 W kg^{-1}
					5-m sprint
					1.15 ± 0.08 s
					30-m sprint
					4.67 ± 0.25 s

(continued)

Table 3.7 (continued)

Study	Team sport	Sample	Group/period	Training intervention	Results
Arcos et al. (2015)		17 (U16)	HIT 8-weeks	2 days per week Running 3 × 4 min (90–95 % HRmax)/3 min active recovery	Preintervention Maximal aerobic speed 16.8 ± 0.9 Countermovement Jump 42.76 ± 4.59 cm Postintervention Maximal aerobic speed 17.1 ± 1.0 Countermovement Jump 42.41 ± 4.76 cm
			SSCG 8-weeks	2 days per week 3 versus 3 and 4 versus 4 3 × 4 min/3 min rest	Preintervention Maximal aerobic speed 17.0 ± 0.8 Countermovement Jump 42.71 ± 2.43 cm Postintervention Maximal aerobic speed 16.9 ± 0.8 Countermovement Jump 41.96 ± 2.76 cm
Buchheit et al. (2009b)	Handball	32 (U16)	HIT 10-weeks	2 days per week 6–12 min of intermittent running for 15 s (95 % V_{IFT}) interspersed with 15 s of passive recovery	Preintervention 10-m sprint 2.02 ± 0.16 s Countermovement Jump 40.70 ± 9.70 cm Repeated Sprint Ability 6.52 ± 0.42 s (mean) 30-15 Intermittent Fitness Test 17.9 ± 1.8 km/h

(continued)

Table 3.7 (continued)

Study	Team sport	Sample	Group/period	Training intervention	Results
					Postintervention
					10-m sprint
					2.00 ± 0.14 s
					Countermovement Jump
					42.00 ± 8.40 cm
					Repeated Sprint Ability
					6.30 ± 0.36 s (mean)
					30-15 Intermittent Fitness Test
					18.9 ± 1.3 km/h
			SSCG 10-weeks	2 days per week 4 versus 4 2 to 4 × 2-min 30 s to 4-min games	Preintervention
					10-m sprint
					2.03 ± 0.10 s
					Countermovement Jump
					38.10 ± 7.90 cm
					Repeated Sprint Ability
					6.48 ± 0.40 s (mean)
					30-15 Intermittent Fitness Test
					18.4 ± 1.5 km/h
					Postintervention
					10-m sprint
					2.00 ± 0.09 s
					Countermovement Jump
					39.30 ± 7.70 cm
					Repeated Sprint Ability
					6.18 ± 0.35 s (mean)
					30-15 Intermittent Fitness Test
					19.6 ± 1.4 km/h

(continued)

Table 3.7 (continued)

Study	Team sport	Sample	Group/period	Training intervention	Results
Iacono et al. (2015)		18 (25.6 years old)	HIT 8-weeks	2 days per week 12–16-min intermittent running of 15-s activity over 40-m shuttles interspersed by 15 s of passive walking recovery	Preintervention YYIRTL1 1297.8 ± 300 m 10-m sprint 1.55 ± 0.08 s 20-m sprint 2.80 ± 0.10 s Agility Test 6.72 ± 0.22 s Bench press 99.4 ± 10.1 kg Countermovement Jump 34.3 ± 5.7 cm 36.9 ± 4.5 cm
					Postintervention YYIRTL1 1601.1 ± 192 m 10-m sprint 1.52 ± 0.07 s 20-m sprint 2.75 ± 0.12 s Agility Test 6.65 ± 0.24 s Bench press 106.2 ± 10.7 kg Countermovement Jump
			SSCG 8-weeks	2 days per week 3 versus 3 5 × 2.25- to 3.10-min bouts of SSCGs with a passive recovery of 1 min between bouts	Preintervention YYIRTL1 1364.4 ± 397 m 10-m sprint 1.54 ± 0.12 s 20-m sprint 2.81 ± 0.12 s

(continued)

Table 3.7 (continued)

Study	Team sport	Sample	Group/period	Training intervention	Results
					Agility Test 6.68 ± 0.25 s Bench press 105.0 ± 22.6 kg Countermovement Jump 36.5 ± 4.5 cm Postintervention YYIRTL1 1723.3 ± 327 m 10-m sprint 1.48 ± 0.12 s 20-m sprint 2.70 ± 0.10 s Agility Test 6.54 ± 0.21 s Bench press 118.2 ± 21.0 kg Countermovement Jump 40.5 ± 4.5 cm
Delextrat and Martinez (2014)	Basketball	18 (U17)	HIIT 6-weeks	2 days per week intermittent running at 95 % of players' V_{IFT} for 15 s, followed by 15 s of active recovery	Preintervention 30-15 Intermittent Fitness Test 17.4 ± 0.7 km/h Repeated Sprint Ability 27.1 ± 1.9 s (total) Upper Body Power 5.91 ± 1.83 m Lower Body Power 10.7 ± 1.3

(continued)

Table 3.7 (continued)

Study	Team sport	Sample	Group/period	Training intervention	Results
			SSCG 6-weeks	2 days per week 2 versus 2 2 × (2 × 3 min 45 s) to 2 × (2 × 4 min 15 s)	Postintervention 30-15 Intermittent Fitness Test 18.0 ± 1.0 km/h Repeated Sprint Ability 27.0 ± 1.8 s (total) Upper Body Power 5.79 ± 1.49 m Lower Body Power 10.9 ± 1.0 m
					Preintervention 30-15 Intermittent Fitness Test 17.2 ± 1.7 km/h Repeated Sprint Ability 27.9 ± 2.4 s (total) Upper Body Power 6.10 ± 1.34 m Lower Body Power 10.7 ± 1.0
					Postintervention 30-15 Intermittent Fitness Test 17.9 ± 1.5 km/h Repeated Sprint Ability 28.7 ± 1.9 s (total) Upper Body Power 6.58 ± 1.29 m Lower Body Power 11.2 ± 0.8 m

It can be observed that multiple conditioning variables (aerobic, anaerobic, speed, and power) are improved based on SSCGs and HIIT training programs. Nevertheless, SSCGs seem to benefit more skill variables in comparison with HIIT (Delextrat and Martínez 2014; Radziminski et al. 2013). For that reason, sports training plans benefit from the application of SSCG programs instead of HIIT running methods. Moreover, the psychological effects of SSCG programs in comparison with running methods include lower perceived exertion and more motivation, pleasure, commitment, and we-stories that increase social bonding (Krustrup et al. 2010). Besides the apparent conditioning and psychological benefits of SSCG programs, it was also found in rugby that the majority of injuries occurs in traditional conditioning activities that involved no skill component other than drill-based activities (Gabbett 2002). For that reason, SSCGs provide a solid contribution to training periodization and the development of conditioning variables.

3.5   Comparing SSCGs with Different Traditional Training Methods

The comparison between SSCGs and HIIT training is the most common. Nevertheless, some other training methods have been compared with SSCGs, mainly to identify the acute responses and training effects on variables of performance associated with the anaerobic system, speed, agility, and power. Table 3.8 summarizes the acute responses in the comparisons between SSCGs and traditional training methods.

A study conducted in soccer that compared speed endurance production and speed endurance maintenance methods between running and SSCGs revealed that generic running drills elicit greater physiological responses; nevertheless, SSCG methods induced greater acceleration and deceleration profiles (Ade et al. 2013). This is the only study found in this review that analyzed acute responses to different training approaches. On the other hand, more studies analyzed the conditioning effects of SSCGs and traditional training programs in soccer and rugby (Chaouachi et al. 2014; Gabbett 2006; Hill-Haas et al. 2009a; Young and Rogers 2014). A summary of the main findings can be observed in Table 3.9.

The first study compared a mixed generic training with an SSCG training program, revealing similar effects of both programs on the improvement of conditioning variables (Hill-Haas et al. 2009a). Effects of both training programmes resulted in differences in VO_{2max} (0–2 % better), distance covered in YYIRTL1 (17–21 % better), 5-m speed (0–1 % faster), 20-m speed (0–1 % worse), and (0–1 % better) (Hill-Haas et al. 2009a). The main findings revealed that both training programs improved the aerobic system (Hill-Haas et al. 2009a). A different approach was used to compare SSCGs with multidirectional sprints' intervention on agility and change of direction (COD) (Chaouachi et al. 2014). COD statistically improved sprinting, agility without ball, and jumping performances in comparison

Table 3.8 Summary of studies comparing the acute responses of SSCGs and multiple running training methods in team sports

Study	Team sport	Sample	Group	Training prescription	Results
Ade et al. (2013)*		21 (Elite U18)	SEP Running	Running 8 × 30 s/120 s recovery	% HRpeak 90.4 % BLa⁻ 13.4 mmol/L Distance covered (Total) 176.4 m Distance covered (Very high speed) 88.1 m Distance covered (Maximal accelerations) 2.5 m Distance covered (Maximum decelerations) 1.1 m
			SEP SSCG	1 versus 1 8 × 30 s/120 s recovery	% HRpeak 89.2 % BLa⁻ 10.1 mmol/L Distance covered (total) 84.5 m Distance covered (Very high speed) 5.6 m Distance covered (Maximal accelerations) 3.3 m Distance covered (Maximum decelerations) 3.3 m

(continued)

Table 3.8 (continued)

Study	Team sport	Sample	Group	Training prescription	Results
			SEM Running	Running 8 × 60 s/60 s recovery	% HRpeak 91.9 %
					BLa⁻ 12.1 mmol/L
					Distance covered (total) 281.7 m
					Distance covered (Very high speed) 73.2 m
					Distance covered (Maximal accelerations) 1.1 m
					Distance covered (Maximum decelerations) 0.2 m
			SEP SSCG	2 versus 2 8 × 60 s/60 s recovery	% HRpeak 91.0 %
					BLa⁻ 6.1 mmol/L
					Distance covered (total) 130.9 m
					Distance covered (Very high speed) 3.9 m
					Distance covered (Maximal accelerations) 4.5 m
					Distance covered (Maximum decelerations) 3.4 m

*SEP Speed endurance production; SEM Speed endurance maintenance

Table 3.9 Summary of studies comparing the effects of SSCGs and multiple running training programs in team sports

Study	Team sport	Sample	Group/period	Training intervention	Results
Hill-Haas et al. (2009a)	Soccer	25 (U15)	Mixed Generic Fitness Training 7-weeks	2 days per week Included aerobic power, prolonged intermittent high-intensity running, sprint training (up to a maximum of 20 m) and repeated sprint training	Preintervention VO_{2max} 60.2 ± 4.6 mL/kg/min YYIRTL1 1764 ± 256 m 5-m sprint 1.16 ± 0.02 s 20-m sprint 3.27 ± 0.06 s Repeated Sprint Ability (Total) 42.2 ± 1.8 s Postintervention VO_{2max} 61.4 ± 3.5 mL/kg/min YYIRTL1 2151 ± 261 m 5-m sprint 1.15 ± 0.05 s 20-m sprint 3.22 ± 0.10 s Repeated Sprint Ability (Total) 42.3 ± 1.5 s

(continued)

Table 3.9 (continued)

Study	Team sport	Sample	Group/period	Training intervention	Results
			SSCG 7-weeks	2 days per week 2 versus 2 to 7 versus 7 2 to 6 bouts 6 to 13 repetitions 1 to 2 min rest	Preintervention VO_{2max} 59.3 ± 4.5 mL/kg/min YYIRTL1 1488 ± 345 m 5-m sprint 1.15 ± 0.05 s 20-m sprint 3.26 ± 0.12 s Repeated Sprint Ability (Total) 42.1 ± 1.1 s Postintervention VO_{2max} 58.9 ± 5.5 mL/kg/min YYIRTL1 1742 ± 362 m 5-m sprint 1.16 ± 0.07 s 20-m sprint 3.24 ± 0.17 s Repeated Sprint Ability (Total) 42.0 ± 1.4 s

(continued)

Table 3.9 (continued)

Study	Team sport	Sample	Group/period	Training intervention	Results
Chaouachi et al. (2014)*		36 (U14)	COD 6-weeks	Included Skipping 10 m, COD 5-0-5 m, Half-T-test 20 m 2 bouts 2 to 4 repetitions	Preintervention 30 m 4.51 ± 0.22 s 20 m 3.15 ± 0.14 s 10 m 1.71 ± 0.07 s Reactive Agility Test Ball Dribbling 2.67 ± 0.18 s Zig-Zag Test 20 m 7.40 ± 0.56 s Countermovement Jump 40.38 ± 2.68 cm Postintervention 30 m 4.33 ± 0.14 s 20 m 3.05 ± 0.13 s 10 m 1.68 ± 0.05 s Reactive Agility Test Ball Dribbling

(continued)

Table 3.9 (continued)

Study	Team sport	Sample	Group/period	Training intervention	Results
			SSCG 6-weeks	1 versus 1, 2 versus 2 and 3 versus 3 1 to 2 bouts 2 to 4 repetitions	2.54 ± 0.16 s Zig-Zag Test 20 m 7.03 ± 0.47 s Countermovement Jump 42.03 ± 2.44 cm Preintervention 30 m 4.53 ± 0.29 s 20 m 3.16 ± 0.18 s 10 m 1.71 ± 0.10 s Reactive Agility Test Ball Dribbling 2.65 ± 0.09 s Zig-Zag Test 20 m 7.70 ± 0.28 s Countermovement Jump 39.43 ± 3.04 cm Postintervention 30 m 4.46 ± 0.23 s 20 m 3.12 ± 0.16 s 10 m 1.69 ± 0.07 s

(continued)

58 3 Small-Sided and Conditioned Games Versus Traditional Training …

Table 3.9 (continued)

Study	Team sport	Sample	Group/period	Training intervention	Results
					Reactive Agility Test Ball Dribbling 2.45 ± 0.11 s Zig-Zag Test 20 m 7.51 ± 0.29 s Countermovement Jump 40.21 ± 2.42 cm
Young and Rogers (2014)		25 (Elite U18)	Change of direction 7-weeks	2 days per week Involved movements requiring short maximum effort sprints around various courses defined by cones or poles resulting in 1–5 changes of direction and/or speed There were 4 activities per session, typically with a slow walk-back recovery between repetitions (15 s)	Preintervention Reactive Agility Test Total Time (s) 2.56 ± 0.10 s Decision Time (s) 0.28 ± 0.07 s Movement Response Time (s) 0.96 ± 0.10 s Postintervention Reactive Agility Test Total Time (s) 2.56 ± 0.09 s Decision Time (s) 0.27 ± 0.05 s Movement Response Time (s) 0.98 ± 0.17 s

(continued)

Table 3.9 (continued)

Study	Team sport	Sample	Group/period	Training intervention	Results
			SSCG 7-weeks	2 days per week 2 versus 2 and 4 versus 4 30–45 s bouts with the same recovery period	Preintervention Reactive Agility Test Total Time (s) 2.64 ± 0.13 s Decision Time (s) 0.35 ± 0.06 s Movement Response Time (s) 0.97 ± 0.08 s Postintervention Reactive Agility Test Total Time (s) 2.54 ± 0.08 s Decision Time (s) 0.24 ± 0.03 s Movement Response Time (s) 0.98 ± 0.09 s

(continued)

Table 3.9 (continued)

Study	Team sport	Sample	Group/period	Training intervention	Results
Gabbett (2006)	Rugby	69 (22.3 years old)	Traditional Conditioning Activities 9-weeks	2 days per week Continuous and intermittent running activities without the ball; Power activities; and Agility activities	Preintervention VO_{2max} 49.6 ± 0.7 mL/kg/min 10 m sprint 1.85 ± 0.01 s 20 m sprint 3.12 ± 0.02 s 40 m sprint 5.60 ± 0.04 s Agility (s) 5.78 ± 0.03 s Vertical Jump 60.1 ± 0.7 cm Postintervention VO_{2max} 52.2 ± 0.5 mL/kg/min 10 m sprint 1.80 ± 0.01 s 20 m sprint 3.12 ± 0.02 s 40 m sprint 5.59 ± 0.03 s Agility (s) 5.74 ± 0.07 s Vertical Jump 57.0 ± 0.8 cm

(continued)

Table 3.9 (continued)

Study	Team sport	Sample	Group/period	Training intervention	Results
			SSCG 9-weeks	2 days per week Multiple versions of SSCGs	Preintervention VO_{2max} 46.6 ± 0.5 mL/kg/min 10 m sprint 1.91 ± 0.02 s 20 m sprint 3.17 ± 0.02 s 40 m sprint 5.64 ± 0.03 s Agility (s) 5.73 ± 0.04 s Vertical Jump 57.5 ± 1.3 cm Postintervention VO_{2max} 48.8 ± 0.4 mL/kg/min 10 m sprint 1.81 ± 0.02 s 20 m sprint 3.07 ± 0.02 s 40 m sprint 5.47 ± 0.03 s Agility (s) 5.70 ± 0.03 s Vertical Jump 60.2 ± 1.1 cm

*COD multidirectional sprints intervention on agility and change of direction

with SSCGs. Nevertheless, SSCG programs statistically improved agility with the ball (Chaouachi et al. 2014). Also in soccer, a study that compared change-of-direction speed training and SSCG programs revealed that SSCGs improve agility performance by enhancing the speed of decision-making rather than movement speed and that change-of-direction training was not effective for developing reactive agility (Young and Rogers 2014). In rugby, SSCG programs induced statistical differences in 10, 20, and 40-m speed as well as in muscular power and maximal aerobic power, whereas traditional conditioning programs improved 10-m speed and maximal aerobic power only (Gabbett 2006).

In short, it was possible to identify in these different studies that SSCGs had multiple effects on conditioning variables, thus overcoming the unique aerobic benefits resulting from games. Speed, agility, and power also benefited from SSCGs based on the specific type of movements produced in a game (such as acceleration/deceleration, jumps, and turns). For that reason, the specificity of training with the game may contribute to a generic improvement in conditioning variables that support performance in competition (Clemente et al. 2014a). Nevertheless, more studies are required to identify whether these changes were influenced by specific training programs or by concurrent training that is promoted by strength and conditioning professionals in the academy.

3.6 Conclusion

This review provides information that can help reduce skepticism about the effects of SSCGs on conditioning variables. The main comparisons between SSCG programs and traditionally run training methods reveal similar effects on aerobic, anaerobic, speed, and power variables. Moreover, the skill improvements and psychological benefits of SSCGs in comparison with traditional methods were also verified in this review, thus demystifying the practical applications of these games for training periodization in team sports.

References

Abdelkrim, N. B., Castagna, C., Fazaa, S. E., & Ati, J. E. (2010). The effect of players' standard and tactical strategy on game demands in men's basketball. *Journal of Strength and Conditioning Research, 24*(10), 2652–2662.

Ade, J., Harley, J., & Bradley, P. (2013). The physiological response, time-motion characteristics and reproducibility of various speed endurance drills in elite youth soccer players: small sided games vs generic running. *British Journal of Sports Medicine, 47*, e4.

Aguiar, M., Botelho, G., Lago, C., Maças, V., & Sampaio, J. (2012). A review on the effects of soccer small-sided games. *Journal of Human Kinetics, 33*, 103–113.

Arcos, A. L., Vázquez, J. S., Martín, J., Lerga, J., Sánchez, F., Villagra, F., & Zulueta, J. J. (2015). Effects of small-sided games vs. interval training in aerobic fitness and physical enjoyment in young elite soccer players. *PLoS ONE, 10*(9), e0137224.

Aroso, J., Rebelo, A. N., & Gomes-Pereira, J. (2004). Physiological impact of selected game-related exercises. *Journal of Sports Sciences, 22*, 522.

Atli, H., Koklu, Y., Alemdaroglu, U., & Koçar, U. (2013). A comparison of heart rate response and frequencies of technical actions between half-court and full-court 3-a-side games in high school female basketball players. *Journal of Strength and Conditioning Research, 27*(2), 352–356.

Berdejo-del-Fresno, D., Moore, R., & W. Laupheimer, M. (2015). Changes in English Futsal Players after a 6-Week Period of Specific Small-Sided Games Training. *American Journal of Sports Science and Medicine, 3*(2), 28–34.

Brandes, M., Heitmann, A., & Müller, L. (2012). Physical responses of different small-sided game formats in elite youth soccer players. *Journal of Strength and Conditioning Research, 26*(5), 1353–1360.

Buchheit, M., Laursen, P. B., Kuhnle, J., Ruch, D., Renaud, C., & Ahmaidi, S. (2009a). Game-based training in young elite handball players. *International Journal of Sports Medicine, 30*(4), 251–258.

Buchheit, M., Lepretre, P. M., Behaegel, A. L, Millet, G. P., Cuvelier, G., & Ahmaidi, S. (2009b). Cardiorespiratory responses during running and sport-specific exercises in handball players. *Journal of Science and Medicine in Sport, 12*(3), 399–405.

Casamichana, D., & Castellano, J. (2010). Time–motion, heart rate, perceptual and motor behaviour demands in small-sides soccer games: Effects of pitch size. *Journal of Sports Sciences, 28*(14), 1615–1623.

Casamichana, D., Castellano, J., & Dellal, A. (2013). Influence of different training regimes on physical and physiological demands during small-sided soccer games. *Journal of Strength and Conditioning Research, 27*(3), 690–697.

Castagna, C., Impellizzeri, F. M., Chaouachi, A., Ben Abdelkrim, N., & Manzi, V. (2011). Physiological responses to ball-drills in regional level male basketball players. *Journal of Sports Sciences, 29*(12), 1329–1336.

Castellano, J., Casamichana, D., & Dellal, A. (2013). Influence of game format and number of players on heart rate responses and physical demands in small-sided soccer games. *Journal of Strength & Conditioning Research, 27*, 1295–1303.

Chaouachi, A., Chtara, M., Hammami, R., Chtara, H., Turki, O., & Castagna, C. (2014). Multidirectional sprints and small-sided games training effect on agility and change of direction abilities in youth soccer. *Journal of Strength & Conditioning Research, 28*(11), 3121–3127.

Clemente, F. M., Lourenço, F. M., & Mendes, R. S. (2014a). Developing aerobic and anaerobic fitness using small-sided soccer games: Methodological proposals. *Strength and Conditioning Journal, 36*(3), 76–87.

Clemente, F. M., Martins, F. M. L., & Mendes, R. S. (2014b). Periodization based on small-sided soccer games. *Strength and Conditioning Journal, 36*(5), 34–43.

Conte, D., Favero, T. G., Niederhausen, M., Capranica, L., & Tessitore, A. (2015a). Effect of different number of players and training regimes on physiological and technical demands of ball-drills in basketball. *Journal of Sports Sciences, ahead-of-p*, 1–7.

Conte, D., Favero, T. G., Niederhausen, M., Capranica, L., & Tessitore, A. (2015b). Physiological and technical demands of no dribble game drill in young basketball players. *Journal of Strength and Conditioning Research, Ahead-of-p*, 1–15.

Davids, K., Araújo, D., Correia, V., & Vilar, L. (2013). How small-sided and conditioned games enhance acquisition of movement and decision-making skills. *Exercise and Sport Sciences Reviews, 41*(3), 154–161.

Delextrat, A., & Kraiem, S. (2013). Heart-rate responses by playing position during ball drills in basketball. *International Journal of Sports Physiology and Performance, 8*, 410–418.

Delextrat, A., & Martínez, A. (2014). Small-sided game training improves aerobic capacity and technical skills in basketball players. *International Journal of Sports Medicine, 35*, 385–391.

Dellal, A., Chamari, K., Pintus, A., Girard, O., Cotte, T., & Keller, D. (2008). Heart rate responses during small-sided games and short intermittent running training in elite soccer players: A comparative study. *Journal of Strength and Conditioning Research, 22*(5), 1449–1457.

Dellal, A., Hill-Haas, S., Lago-Penas, C., & Chamari, K. (2011). Small-sided games in soccer: amateur vs. professional players' physiological responses, physical, and technical activities. The. *Journal of Strength & Conditioning Research, 25*(9), 2371–2381.

Dellal, A., Varliette, C., Owen, A., Chirico, E. N., & Pialoux, V. (2012). Small-sided games versus interval training in amateur soccer players: Effects on the aerobic capacity and the ability to perform intermittent exercises with changes of direction. *Journal of Strength and Conditioning Research, 26*(10), 2712–2720.

Fanchini, M., Azzalin, A., Castagna, C., Schena, F., McCall, A., & Impellizzeri, F. M. (2011). Effect of bout duration on exercise intensity and technical performance of small-sided games in soccer. *Journal of Strength and Conditioning Research/National Strength & Conditioning Association, 25*(2), 453–458.

Foster, C. D., Twist, C., Lamb, K. L., & Nicholas, C. W. (2010). Heart rate responses to small-sided games among elite junior rugby league players. *Journal of Strength and Conditioning Research, 24*(4), 906–911.

Gabbett, T. J. (2002). Training injuries in rugby league: An evaluation of skill-based conditioning games. *Journal of Strength & Conditioning Research, 16*(2), 236–241.

Gabbett, T. J. (2006). Skill-based conditioning games as an alternative to traditional conditioning for rugby league players. *The Journal of Strength and Conditioning Research, 20*(2), 309. doi:10.1519/R-17655.1.

Gabbett, T., Jenkins, D., & Abernethy, B. (2009). Game-based training for improving skill and physical fitness in team sport athletes. *International Journal of Sports Science and Coaching, 4*(2), 273–283.

Gracia, F., García, J., Cañadas, M., & Ibáñez, S. J. (2014). Heart rate differences in small-sided games in formative basketball. *Journal of Sport Science, 10*(1), 23–30.

Halouani, J., Chtourou, H., Dellal, A., Chaouachi, A., & Chamari, K. (2014a, April). Physiological responses according to rules changes during 3 vs. 3 small-sided games in youth soccer players: Stop-ball vs. small-goals rules. *Journal of Sports Sciences*, 37–41.

Halouani, J., Chtourou, H., Gabbett, T., Chaouachi, A., & Chamari, K. (2014b). Small-sided games in team sports training: Brief review. *Journal of Strength and Conditioning Research, 28*(12), 3594–3618.

Hill-Haas, S. V., Coutts, a J, Rowsell, G. J., & Dawson, B. T. (2009a). Generic versus small-sided game training in soccer. *International Journal of Sports Medicine, 30*(9), 636–642. doi:10.1055/s-0029-1220730.

Hill-haas, S. V., Rowsell, G. J., Dawson, B. T., & Coutts, A. J. (2009b). Acute physiological responses and time-motion characteristics of two small-sided training regimes in youth soccer players. *Journal of Strength and Conditioning Research, 23*(1), 111–115.

Hill-Haas, S. V., Dawson, B., Impellizzeri, F. M., & Coutts, A. J. (2011). Physiology of small-sided games training in football. *Sports Medicine, 41*(3), 199–220.

Hoffmann, J. J, Jr., Reed, J. P., Leiting, K., Chiang, C. Y., & Stone, M. H. (2014). Repeated sprints, high intensity interval training, small sided games: Theory and application to field sports. *International Journal of Sports Physiology and Performance, 9*(2), 352–357.

Iacono, A. D., Eliakim, A., & Meckel, Y. (2015). Improving fitness of elite handball players: Small-sided games vs. high-intensity intermittent training. *Journal of Strength & Conditioning Research, 29*(3), 835–843.

Impellizzeri, F. M., Marcora, S. M., Castagna, C., Reilly, T., Sassi, A., Iaia, F. M., & Rampinini, E. (2006a). Physiological and performance effects of generic versus specific aerobic training in soccer players. *International Journal of Sports Medicine, 27*(6), 483–92.

Impellizzeri, F. M., Marcora, S. M., Castagna, C., Reilly, T., Sassi, A., Iaia, F. M., & Rampinini, E. (2006b). Physiological and performance effects of generic versus specific aerobic training in soccer players. *International Journal of Sports Medicine, 27*, 483–492.

Impellizzeri, F. M., Rampinini, E., & Marcora, S. M. (2005). Physiological assessment of aerobic training in soccer. *Journal of Sports Sciences, 23*(6), 583–592.

Jake, N., Tsui, M., Smith, A., Carling, C., Chan, G., & Wong, D. (2012). The effects of man-marking on work intensity in small-sided soccer games. *Journal of Sports Science and Medicine, 11*, 109–114.

Jastrzebski, Z., Barnat, W., Dargiewicz, R., Jaskulska, E., Szwarc, A., & Radziminski, L. (2014). Effect of in-season generic and soccer-specific high-intensity interval training in young soccer players. *International Journal of Sports Science & Coaching, 9*(5), 1169–1179.

Kelly, D. M., & Drust, B. (2009). The effect of pitch dimensions on heart rate responses and technical demands of small-sided soccer games in elite players. *Journal of Science and Medicine in Sport, 12*(4), 475–479.

Kennett, D. C., Kempton, T., & Coutts, A. J. (2012). Factors affecting exercise intensity in rugby-specific small-sided games. *Journal of Strength and Conditioning Research, 26*(8), 2037–2042.

Klusemann, M. J., Pyne, D. B., Foster, C., & Drinkwater, E. J. (2012). Optimising technical skills and physical loading in small-sided basketball games. *Journal of Sports Sciences, 30*(14), 1463–1471.

Köklü, Y. (2012). A comparison of physiological responses to various intermittent and continuous small-sided games in young soccer players. *Journal of Human Kinetics, 31*, 89–96.

Köklü, Y., Asçi, A., Koçak, F. Ü., Alemdaroglu, U., & Dündar, U. (2011). Comparison of the physiological responses to different small-sided games in elite young soccer players. *The Journal of Strength & Conditioning Research, 25*(6), 1522–1528.

Krustrup, P., Dvorak, J., Junge, A., & Bangsbo, J. (2010). Executive summary: the health and fitness benefits of regular participation in small-sided football games. *Scandinavian Journal of Medicine and Science in Sports, 20*(Suppl 1), 132–135.

Little, T. (2009). Optimizing the use of soccer drills for physiological development. *Strength and Conditioning Journal, 31*(3), 67–74.

Little, T., & Williams, A. G. (2006). Suitability of soccer training drills for endurance training. *Journal of Strength and Conditioning Research, 20*(2), 316–319.

Mallo, J., & Navarro, E. (2008). Physical load imposed on soccer players during small-sided training games. *The Journal of Sports Medicine and Physical Fitness, 48*, 166–171.

McCormick, B. T., Hannon, J. C., Newton, M., Shultz, B., Miller, N., & Young, W. (2012). Comparison of physical activity in small-sided basketball games versus full-sided games. *International Journal of Sports Science and Coaching, 7*(4), 689–698.

Owen, A. L., Wong, D. P., McKenna, M., & Dellal, A. (2011). Heart rate responses and technical comparison between small- vs. large-sided games in elite professional soccer. *Journal of Strength and Conditioning Research, 25*(8), 2104–2110.

Owen, A. L., Wong, D. P., Paul, D., & Dellal, A. (2012). Effects of a periodized small-sided game training intervention on physical performance in elite professional soccer. *Journal of Strength and Conditioning Research, 26*(10), 2748–2754.

Owen, A., Twist, C., & Ford, P. (2004). Small-sided games: the physiological and technical effect of altering field size and player numbers. *Insight, 7*, 50–53.

Radziminski, L., Rompa, P., Barnat, W., Dargiewicz, R., & Jastrzebski, Z. (2013). A comparison of the physiological and technical effects of high-intensity running and small-sided games in young soccer players. *International Journal of Sports Science & Coaching, 8*(3), 455–465.

Rampinini, E., Impellizzeri, F. M., Castagna, C., Abt, G., Chamari, K., Sassi, A., & Marcora, S. M. (2007). Factors influencing physiological responses to small-sided soccer games. *Journal of Sports Sciences, 25*(6), 659–666.

Reilly, T., & White, C. (2004). Small-sided games as an alternative to interval-training for soccer players [abstract]. *Journal of Sports Sciences, 22*(6), 559.

Sampaio, J., Abrantes, C., & Leite, N. (2009). Power, heart rate and perceived exertion responses to 3x3 and 4x4 basketball small-sided games. *Revista de Psicología Del Deporte, 18*(3), 463–467.

Sampaio, J., Garcia, G., Macas, V., Ibànez, S., Abrantes, C., & Caixinha, P. (2007). Heart rate and perceptual responses to 2x2 and 3x3 small-sided youth soccer games. *Journal of Sports Science & Medicine, 6*(10), 121–122.

Sampaio, J., Leser, R., Baca, A., Calleja-gonzalez, J., Coutinho, D., Gonçalves, B., & Leite, N. (2015). Defensive pressure affects basketball technical actions but not the time-motion variables. *Journal of Sport and Health Science, Ahead-of-print.*, doi:10.1016/j.jshs.2015.01.011.

Sassi, R., Reilly, T., & Impellizzeri, F. M. (2004). A comparison of small-sided games and interval training in elite professional soccer players [abstract]. *Journal of Sports Sciences, 22*, 562.

Seitz, L. B., Rivière, M., Villarreal, E. S., & Haff, G. G. (2014). The athletic performance of elite rugby league players is improved after an 8-week small-sided game training intervention. *Journal of Strength & Conditioning Research, 28*(4), 971–975.

Williams, K., & Owen, A. (2007). The impact of player numbers on the physiological responses to small sided games. *Journal of Sports Science & Medicine, Suppl, 10*, 99–102.

Young, W., & Rogers, N. (2014). Effects of small-sided game and change-of-direction training on reactive agility and change-of-direction speed. *Journal of Sports Sciences, 32*(4), 307–314.

Chapter 4
Acute Effects of Different Formats of the Game

Abstract The number of players that participates in smaller versions of the game influences the training load. This variable has been well investigated in the specific literature about small-sided and conditioned games and for that reason, will be presented in first place. In this chapter will be analyzed the internal and external load imposed by different formats of the game and the specific effects in technical actions and tactical behavior of the players. The aim of this chapter is to summarize the most pertinent information about each format of the game (from one versus one to many versus many) and provide coaches the knowledge that can help them to choose the most adequate formats for their specific training goals.

Keywords Training load · Format of the game · Number of players · Small-sided and conditioned games · SSG · Drill-based exercises · Soccer · Football · Sports training

4.1 Introduction

The format of the game (number of players on each team) in a small-sided and conditioned game (SSCG) can be altered to regulate the intensity of the training mode (Hill-Haas et al. 2011). One of the main concerns that this condition implies during researches is to keep the same area per players (Clemente et al. 2014b). Increasing the format and keeping the same field will naturally reduce the area per player and another variable will emerge in the equation: the size of the field. Nevertheless, the aim of this chapter is only to focus on the physiological, physical, and technical/tactical changes that result from the change in the format.

This chapter will summarize the studies conducted about this topic. The structure will try to summarize the scientific evidences per each format, thus providing to the reader the opportunity to easily identify the general effects of each format. Based on that we will have the opportunity to decide about the most adequate format for each type of period of week or of the training session.

© The Author(s) 2016
F.M. Clemente, *Small-Sided and Conditioned Games in Soccer Training*,
SpringerBriefs in Applied Sciences and Technology,
DOI 10.1007/978-981-10-0880-1_4

4.2   1 Versus 1 Format

The 1 versus 1 format can be called by duel. This extreme SSCG leads to very high levels of effort and for that reason must be treated as a specific drill for anaerobic training (Clemente et al. 2014a; Little 2009). The research in this specific format is not so large as comparing with bigger formats. Nevertheless, the majority of the studies prescribed 1 to 3 min of exercise, with a ratio 1:1 of work-to-rest (Clemente et al. 2014a; Little 2009). Two to four bouts for a total volume of 16 min (maximum) is recommended for this kind of task (Clemente et al. 2014a) (Table 4.1).

The few studies that analyzed this format (see Table 4.2) revealed a blood lactate concentration of 9.4 (greater than lactate threshold) and intensities ~ 86 % HRmax (Köklü et al. 2011; Owen et al. 2004; Williams and Owen 2007). No study analyzed the time–motion profile of players in this format. The unique technical analysis carried out on this format revealed a bigger tendency to do dribbles, turns, and headers in comparison with bigger formats (Owen et al. 2004).

Table 4.1 Acute physiological effects during 1 versus 1 format

Study	Participants	SF	Regimen	HR	BLa^{-1}
Owen et al. (2004)	13 (U17)	10 × 5	1 × 3/12 min rest	176 bpm	–
Owen et al. (2004)	13 (U17)	15 × 10	1 × 3/12 min rest	181 bpm	–
Owen et al. (2004)	13 (U17)	20 × 15	1 × 3/12 min rest	182 bpm	–
Williams and Owen (2007)	9 (U17)	20 × 15	–	183 bpm	–
Dellal et al. (2008)	10 (elite)	10 × 10	4 × 1 min, 30 s/1 min, 30 s rest	77.6 HRres	–
Köklü et al. (2011)	16 (U16)	6 × 18	6 × 1 min/2 min rest	168.6 bpm 86.1 % HRmax	9.4

SF Size of the field (m); *HR* Heart rate; *BLa^{-1}* Blood lactate concentration (mmol/L)

Table 4.2 Technical performance during 1 versus 1 format

Study	Participants	SF	Regimen	Dribble	Turn	Header
Owen et al. (2004)	13 (U17)	10 × 5 15 × 10 20 × 15	1 × 3/12 min rest	3 per player	4 per player	1 per player

4.3   2 Versus 2 Format

Similarly to duels, 2 versus 2 format is a highly demanding task. The studies that analyzed this drill (see Table 4.3) identified values between 3.4 and 8.1 of blood lactate concentrations, thus suggesting values in the lactate threshold (Aroso et al. 2004; Köklü et al. 2011). The intensity values vary between 80.1 and 93.3 % HRmax, thus confirming that glycolytic system highly participate during these games (Dellal et al. 2011b; Little and Williams 2007). Duration of the task may vary between 1 min and 30 s and the 3 min in 2–4 bouts with a work-to-rest ratio of 1:1 for a total volume of 16 min (Clemente et al. 2014a; Little 2009).

The time–motion analysis carried out in this format (see Table 4.4) revealed that players cover 100–144 m per min in the majority of time in walk or jogging mode (Hill-Haas et al. 2009; Dellal et al. 2011a, b). Only during ∼3.5 % of the time can be observed sprints and very fast runs.

During 2 versus 2 format it was possible observe an accuracy between 62 and 66.4 % of the passes and a tendency to perform 12–13 duels per min, thus suggesting an interesting opportunity to develop the basic skills of soccer (Table 4.5).

4.4   3 Versus 3 Format

As possible to observe in Table 4.6, 3 versus 3 format keeps very high intensity (87–94 % HRmax) without a great blood lactate concentration (3–7.5 mmol/L). This format is one of the most studied in the field of SSCGs, maybe by their limited position between extreme SSCGs (1 vs. 1 or 2 vs. 2) and the small-sided games with greater number of players. In the majority of these studies the prescription was 3–6 min with 2–3 bouts and a work-to-rest ratio of 1:0.5 (Clemente et al. 2014a; Little 2009).

The studies that analyzed the time–motion profile during 3 versus 3 format (see Table 4.7) revealed that players cover 115–160 m per min (Dellal et al. 2011a, b; Aguiar et al. 2013). In the study conducted in elite players (Dellal et al. 2011a, b) it was found that 35 % of the distance covered is made in high intensity or sprint, thus a greater percentage than in 2 versus 2 format. This can be justified by the increase of opportunity to create lines of pass far away of the player with possession of the ball.

The studies (see Table 4.8) revealed that in 3 versus 3 format each player performs ∼7 contacts in the ball per minute. Moreover, 5–12 passes are performed per each minute and there are ∼9 duels per minute. There are fewer duels in 3 versus 3 that in comparison with 2 versus 2 format. For that reason, extreme SSCGs may be better to increase the individual participation and 3 versus 3 may be better to introduce some collective issues such as generate lines of pass or develop the tactical perception.

4 Acute Effects of Different Formats of the Game

Table 4.3 Acute physiological effects during 2 versus 2 format

Study	Participants	SF	Regimen	HR	BLa^{-1}	RPE
Aroso et al. (2004)	14 (U16)	30 × 20	3 × 1 min, 30 s/1 min, 30 s rest	84.0 % HRmax	8.1	16.2 [0–20 scale]
Owen et al. (2004)	13 (U17)	15 × 10	1 × 3/12 min rest	172 bpm	–	–
Owen et al. (2004)	13 (U17)	20 × 15	1 × 3/12 min rest	179 bpm	–	–
Owen et al. (2004)	13 (U17)	25 × 20	1 × 3/12 min rest	180 bpm	–	–
Sampaio et al. (2007)[a]	8 (U15)	30 × 20	2 × 1 min, 30 s/1 min, 30 s rest	83.7 % HRmax	–	15.5 [0–20 scale]
Williams and Owen (2007)	9 (U17)	20 × 15	–	179 bpm	–	–
Williams and Owen (2007)	9 (U17)	25 × 20	–	180 bpm	–	–
Little and Williams (2007)	28 (elite)	30 × 20	4 × 2 min/2 min rest	88.8 % HRmax	–	16.2 [0–20 scale]
Dellal et al. (2008)	10 (elite)	20 × 20	6 × 2 min, 30 s/2 min, 30 s rest	80.1 % HRmax	–	–
Hill-Haas et al. (2009)	16 (U17)	28 × 21	24 min	89 % HRmax	6.7	13.1 [0–20 scale]
Köklü et al. (2011)	16 (U16)	12 × 24	6 × 2 min/2 min rest	172.3 bpm / 88 % HRmax	8.0	–
Dellal et al. (2011a, b)[b]	20 (elite)	20 × 15	4 × 2 min/3 min rest	182 bpm / 90 % HRmax	3.4	7.6 [0-10 scale]
Dellal et al. (2011a, b)	20 (elite) 20 (amateurs)	20 × 15	4 × 2 min/3 min rest	90.0 % HRmax (elite) 91.6 % HRmax (amateur)	3.5 (elite) 4.1 (amateurs)	7.7 [0-10 scale] (elite) 8.0 [0-10 scale] (amateurs)
Brandes et al. (2012)	17 (U15)	28 × 21	3 × 4 min	93.3 % HRmax	–	–
Aguiar et al. (2013)	10 (U18)	150 m² per player	3 × 6 min/1 min rest	87.46 % HRmax	–	–
Clemente et al. (2014b)[c]	10 (amateurs)	19 × 19	3 × 5 min/3 min rest	75.98 % HRres	–	–

SF Size of the field (m); *HR* Heart rate; BLa^{-1} Blood lactate concentration (mmol/L); *RPE* Rated of perceived exertion

[a]HR values in 2 versus 2 with verbal encouragement during task

[b]Values of free play

[c]2 versus 2+2 floaters—values of task with one small goal

Table 4.4 Time–motion analysis during 2 versus 2 format

Study	Participants	SF	Regimen	TD	TD 0–6.9	TD 7.0–12.9	TD 13.0–17.9	TD > 18
Hill-Haas et al. (2009)	16 (U17)	28 × 21	24 min	2574	1176	933	411	44
Dellal et al. (2011a, b)[a]	20 (elite) 20 (amateurs)	20 × 15	4 × 2 min/3 min rest	1157.7 (elite) 1086.7 (amateurs)	-	-	245.5 (elite) 225.7 (amateurs)	177.6 (elite) 160.2 (amateurs)
Aguiar et al. (2013)	10 (U18)	150 m² per player	3 × 6 min/_ min rest	598.97	291.84	232.6	64.04	10.48
Clemente et al. (2014a)[b]	10 (amateurs)	19 × 19	3 × 5 min/5 min rest	240	-	-	-	-

TD Total distance (m); *TD* 0–6.9 Total distance at 0–6.9 km h^{-1}; *TD* 7.0–12.9 Total distance at 7.0–12.9 km h^{-1}; *TD* 13.0–17.9 Total distance at 13.0–17.9 km h^{-1}; *TD* > 18 Total distance at > 18 km h^{-1}

[a]Values of free play

[b]2 versus 2+2 floaters—values of task with one small goal

Table 4.5 Technical performance during 2 versus 2 format

Study	Participants	SF	Regimen	Indicator		Indicator		Indicator	
Owen et al. (2004)	13 (U17)	15 × 10 20 × 15 25 × 20	1 × 3/12 min rest	11 passes per player		7 receives per player		3 dribbles per player	
Dellal et al. (2011a, b)	20 (elite) 20 (amateurs)	20 × 15	4 × 2 min/3 min rest	49.9 possession per min (elite) 41.6 (amateurs		66.4 % successful passes (elite) 62.0 % (amateurs)		26.1 duels (elite) 25 (amateurs)	
Clemente et al. (2014a)[a]	10 (amateurs)	19 × 19	3 × 5 min/3 min rest	17.50 volume of play		0.04 efficiency index		9.18 performance score	

[a]2 versus 2+2 floaters—values of task with one small goal

Table 4.6 Acute physiological effects during 3 versus 3 format

Study	Participants	SF	Regimen	HR	BLa^{-1}	RPE
Aroso et al. (2004)	14 (U16)	30 × 20	3 × 4 min/1 min, 30 s rest	87 % HRmax	4.9	14.5 [0–20 scale]
Owen et al. (2004)	13 (U17)	15 × 20	1 × 3/12 min rest	167 bpm	–	–
Owen et al.(2004)	13 (U17)	20 × 25	1 × 3/12 min rest	167 bpm	–	–
Owen et al. (2004)	13 (U17)	25 × 30	1 × 3/12 min rest	173 bpm	–	–
Sampaio et al. (2007)[a]	8 (U15)	30 × 20	2 × 3 min/1 min, 30 s rest	162.2 bpm 80.8 % HRmax	–	15.8 [0–20 scale]
Williams and Owen (2007)	9 (U17)	20 × 15	–	164 bpm	–	–
Williams and Owen (2007)	9 (U17)	25 × 20	–	166 bpm	–	–
Williams and Owen (2007)	9 (U17)	30 × 25	–	171 bpm	–	–
Little and Williams (2007)	28 (elite)	43 × 25	4 × 3 min, 30 s/1 min, 30 s rest	91 % HRmax	–	15.5 [0–20 scale]
Rampinini et al. (2007)[b]	20 (Amateurs)	12 × 20	3 × 4 min/3 min rest	89.5 % HRmax	6.0	8.1 [0–10 scale]
Rampinini et al. (2007)[b]	20 (Amateurs)	15 × 25	3 × 4 min/3 min rest	90.5 % HRmax	6.3	8.4 [0–10 scale]
Rampinini et al. (2007)[b]	20 (Amateurs)	18 × 30	3 × 4 min/3 min rest	90.9 % HRmax	6.5	8.5 [0–10 scale]
Katis and Kellis (2009)	34 (U14)	15 × 25	10 × 4 min/3 min rest	87.6 % HRmax	–	–
Dellal et al. (2011a)[c]	20 (elite)	25 × 18	4 × 3 min/3 min rest	181 bpm 89.6 % HRmax	3.0	7.5 [0–10 scale]
Owen et al. (2011)	15 (elite)	30 × 25	3 × 5 min/4 min rest	90 % HRmax	–	–

(continued)

Table 4.6 (continued)

Study	Participants	SF	Regimen	HR	BLa^{-1}	RPE
Köklü et al. (2011)	16 (U16)	18 × 30	6 × 3 min/2 min rest	181.7 bpm / 92.8 % HRmax	7.5	–
Da silva et al. (2011)	17 (U15)	30 × 30	3 × 4 min/3 min rest	89.8 % HRmax	–	–
Dellal et al. (2011b)[c]	20 (elite) / 20 (amateurs)	25 × 18	4 × 3 min/3 min rest	89.6 % HRmax (elite) / 89.5 % HRmax (amateurs)	3.1 (elite) / 3.7 (amateurs)	7.5 [0–10 scale] (elite) / 7.7 [0–10 scale] (amateurs)
Brandes et al. (2012)	17 (U15)	34 × 26	3 × 5 min	91.5 % HRmax	3.4	–
Aguiar et al. (2013)	10 (U18)	150 m^2 per player	3 × 6 min/1 min rest	89.56 % HRmax	–	–
Castellano et al. (2013)	14 (semi-professional)	43 × 30	3 × 3 min/5 min rest	93.8 % HRmax	–	–
Clemente et al. (2014a)[d]	10 (amateurs)	23 × 23	3 × 5 min/3 min rest	81.98 % HRres	–	–

SF Size of the field (m); *HR* Heart rate; *BLa^{-1}* Blood lactate concentration (mmol/L); *RPE* Rated of perceived exertion

[a]HR values in 3 versus 3 with verbal encouragement during task

[b]HR values in 3 versus 3 with verbal encouragement during task

[c]Values of free play

[d]2 versus 2+2 floaters—values of task with one small goal

Table 4.7 Time–motion analysis during 3 versus 3 format

Study	Participants	SF	Regimen	TD	TD 0–6.9	TD 7.0–12.9	TD 13.0–17.9	TD > 18
Dellal et al. (2011a, b)[a]	20 (elite)	25 × 18	4 × 3 min/3 min rest	2013.9	–	–	422.4	315.6
Dellal et al. (2011a, b)[a]	20 (elite) 20 (amateurs)	25 × 18	4 × 3 min/3 min rest	2014.00 (elite) 1861 (amateurs)	–	–	422.5 (elite) 383.9 (amateurs)	315.16 (elite) 272.2 (amateurs)
Aguiar et al. (2013)	10 (U18)	150 m^2 per player	3 × 6 min/1 min rest	685.71	278.44	267.44	110.42	29.42
Castellano et al. (2013)	14 (semi-professional)	43 × 30	3 × 3 min/5 min rest	506.6	170.0	173.3	56.3	12.0
Clemente et al. (2014a)[b]	10 (amateurs)	23 × 23	3 × 5 min/3 min rest	240	–	–	–	–

TD Total distance (m); *TD* 0–6.9 Total distance at 0–6.9 km h^{-1}; *TD* 7.0–12.9 Total distance at 7.0–12.9 km h^{-1}; *TD* 13.0–17.9 Total distance at 13.0–17.9 Total distance at 13.0–17.9 km h^{-1}; *TD* > 18 Total distance at > 18 km h^{-1}
[a]Values of free play
[b]2 versus 2 floaters—values of task with one small goal

Table 4.8 Technical performance during 3 versus 3 format

Study	Participants	SF	Regimen	Indicator	Indicator	Indicator	Indicator
Owen et al. (2004)	13 (U17)	15 × 20 20 × 25 25 × 30	1 × 3/12 min rest	8 passes per player	5 receives per player	2 dribbles per player	
Katis and Kellis (2009)	34 (U14)	15 × 25	10 × 4 min/3 min rest	48 short passes	7 long passes	8 dribbles	
Dellal et al. (2011a, b)[a]	20 (elite)	25 × 18	4 × 3 min/3 min rest	26.8 number of duels	71 % successful passes	14.3 balls lost	
Dellal et al. (2011a, b)[a]	20 (elite) 20 (amateurs)	25 × 18	4 × 3 min/3 min rest	26.8 duels (elite) 21.2 (amateurs)	71.7 % successful passes (elite) 70.0 % (amateurs)	41.7 possessions (elite) 37.4 (amateurs)	
Owen et al. (2011)	15 (elite)	30 × 25	3 × 5 min/4 min rest	111 ball contacts per player	193 passes	185 receives	
Da Silva et al. (2011)	17 (U15)	30 × 30	3 × 4 min/3 min rest	31 involvements with ball	19 passes	4 dribbles	
Clemente et al. (2014a)[b]	10 (amateurs)	23 × 23	3 × 5 min/3 min rest	12.25 volume of play	0.30 efficiency index	6.42 performance score	

[a]Values of free play
[b]2 versus 2+2 floaters—values of task with one small goal

4.5 4 Versus 4 Format

The 4 versus 4 format can be classified as a SSCG with aerobic and anaerobic characteristics. The values of intensity are between 70 and 90 % HRmax, nevertheless the majority of the studies are between 84 % and 89 % of HRmax (see Table 4.9). For that reason, this format can be appropriated to develop high-intensity aerobic training. The blood lactate concentrations are between 3 and 7 mmol/L, thus lightly above the lactate threshold. Duration of 4–6 min with 3–4 bouts and a work-to-rest ratio of 1:0.5 for a maximum volume of 30 min are the recommendations to prescribe this format during training sessions (Clemente et al. 2014a; Little 2009).

The majority of the studies that analyzed the time–motion profile during this format revealed that players cover ∼115 m per min (see Table 4.10). The studies found that 12–19 % of the distance is covered in high-intensity running or in sprint, thus less than in 3 versus 3 format. Such evidence may justify the smaller acute effects in heart rate responses and blood lactate concentrations.

In Table 4.11 it can be found the studies that analyzed the technical performance during 4 versus 4 format. Studies revealed that ∼13 passes per min are made

Table 4.9 Acute physiological effects during 4 versus 4 format

Study	Participants	SF	Regimen	HR	BLa^{-1}	RPE
Aroso et al. (2004)	14 (U16)	30 × 20	3 × 6 min/1 min, 30 s rest	70 % HRmax	2.6	13.3 [0–20 scale]
Owen et al. (2004)	13 (U17)	20 × 25	1 × 3/12 min rest	147 bpm	–	–
Owen et al. (2004)	13 (U17)	25 × 30	1 × 3/12 min rest	160 bpm	–	–
Owen et al. (2004)	13 (U17)	30 × 35	1 × 3/12 min rest	158 bpm	–	–
Williams and Owen (2007)	9 (U17)	25 × 20	–	152 bpm	–	–
Williams and Owen (2007)	9 (U17)	30 × 25	–	165 bpm	–	–
Little and Williams (2007)	28 (elite)	40 × 30	4 × 4 min/2 min rest	90.2 % HRmax	–	15.5 [0–20 scale]
Rampinini et al. (2007)[b]	20 (Amateurs)	16 × 24	3 × 4 min/3 min rest	88.7 % HRmax	5.3	7.6 [0–10 scale]
Rampinini et al. (2007)[a]	20 (Amateurs)	20 × 30	3 × 4 min/3 min rest	89.4 % HRmax	5.5	7.9 [0–10 scale]
Rampinini et al. (2007)[a]	20 (Amateurs)	24 × 36	3 × 4 min/3 min rest	89.7 % HRmax	6.0	8.1 [0–10 scale]

(continued)

Table 4.9 (continued)

Study	Participants	SF	Regimen	HR	BLa^{-1}	RPE
Jones and Drust (2007)	8 (elite)	30 × 25	10 min	175 bpm	–	–
Hill-Haas et al. (2009)	16 (U17)	400 × 30	24 min	85 % HRmax	4.7	12.2 [0-20 scale]
Da Silva et al. (2011)	17 (U15)	30 × 30	3 × 4 min/3 min rest	89.8 % HRmax	–	–
Köklü et al. (2011)	16 (U16)	24 × 36	6 × 4 min/2 min rest	179.3 bpm 91.5 % HRmax	7.2	–
Dellal et al. (2011a, b)^{b}	20 (elite) 20 (amateurs)	30 × 20	4 × 4 min/3 min rest	84.7 % HRmax (elite) 85.1 % HRmax (amateurs)	2.8 (elite) 3.0 (amateurs)	7.3 [0–10 scale] (elite) 7.6 [0–10 scale] (amateurs)
Brandes et al. (2012)	17 (U15)	40 × 30	3 × 6 min	89.7 % HRmax	4.2	–
Aguiar et al. (2013)	10 (U18)	150 m^2 per player	3 × 6 min/1 min rest	85.91 % HRmax	–	–
Clemente et al. (2014a)^{c}	10 (amateurs)	27 × 27	3 × 5 min/3 min rest	83.61 % HRres	–	–

SF Size of the field (m); *HR* Heart rate; *BLa*$^{-1}$ Blood lactate concentration (mmol/L); *RPE* Rated of perceived exertion
[a]HR values in 4 versus 4 with verbal encouragement during task
[b]Values of free play
[c]2 versus 2+2 floaters—values of task with one small goal

during this format and the accuracy is greater than 73 %. Three to four individual ball contacts are performed per minute. Therefore, there is an increase of passes per minute in comparison with 3 versus 3 and a decrease in individual ball contacts.

4.6   5 Versus 5 Format

Based on the classification of Owen et al. (2014), 5 versus 5 format can be called by medium-sided game. The heart rate responses are between 85 and 93 % of HRmax in this format. Blood lactate concentration varies between 5 and 5.8 mmol/L. The internal load influenced by this format can be described as similar with 4 versus 4 format. For that reason, this can be used to high-intensity aerobic training. Repetitions of 4–6 min with 3–4 bouts and a work-to-rest ratio of 1:0.5 for a maximum volume of 30 min are recommended (Clemente et al. 2014a; Little 2009) (Table 4.12).

Table 4.10 Time–motion analysis during 4 versus 4 format

Study	Participants	SF	Regimen	TD	TD 0–6.9	TD 7.0–12.9	TD 13.0–17.9	TD > 18
Jones and Drust (2007)	8 (elite)	30 × 25	10 min	778	–	–	–	–
Hill-Haas et al. (2009)	16 (U17)	400 × 30	24 min	2650	1128	1041	436	65
Dellal et al. (2011a, b)[a]	20 (elite) 20 (amateurs)	30 × 20	4 × 4 min/3 min rest	2663.7 (elite) 2419.8 (amateurs)	–	–	482.7 (elite) 480.4 (amateurs)	381.8 (elite) 363.0 (amateurs)
Aguiar et al. (2013)	10 (U18)	150 m² per player	3 × 6 min/1 min rest	682.14	272.74	292.11	96.2	21.09
Owen et al. (2014)	10 (elite)	30 × 25	3 × 5 min/3 min rest	1709	534	963	200	9
Clemente et al. (2014a)[b]	10 (amateurs)	27 × 27	3 × 5 min/3 min rest	290	–	–	–	–

TD Total distance (m); TD 0–6.9 Total distance at 0–6.9 km h^{-1}; TD 7.0–12.9 Total distance at 7.0–12.9 km h^{-1}; TD 13.0–17.9 Total distance at 13.0–17.9 km h^{-1}; TD > 18 Total distance at > 18 km h^{-1}

[a]Values of free play

[b]2 versus 2+2 floaters—values of task with one small goal

Table 4.11 Technical performance during 4 versus 4 format

Study	Participants	SF	Regimen	Indicator	Indicator	Indicator
Owen et al. (2004)	13 (U17)	20 × 25 25 × 30 30 × 35	1 × 3/12 min rest	7 passes per player	4 receives per player	1 dribble per player
Jones and Drust (2007)	8 (elite)	30 × 25	10 min	36 individual ball contacts	–	–
Da Silva et al. (2011)	17 (U15)	30 × 30	3 × 4 min/3 min rest	32 involvements with the ball	20 passes	2 dribbles
Dellal et al. (2011a, b)[a]	20 (elite) 20 (amateurs)	30 × 20	4 × 4 min/3 min rest	25.1 duels (elite) 21.8 (amateurs)	73.5 % successful passes (elite) 70.7 % (amateurs)	31.5 possessions (elite) 35.6 (amateurs)
Owen et al. (2014)	10 (elite)	30 × 25	3 × 5 min/3 min rest	199 passes	166.5 receives	31 dribbles
Clemente et al. (2014a)[b]	10 (amateurs)	27 × 27	3 × 5 min/3 min rest	7.70 volume of play	0.09 efficiency index	4.76 performance score

[a]Values of free play

Table 4.12 Acute physiological effects during 5 versus 5 format

Study	Participants	SF	Regimen	HR	BLa^{-1}	RPE
Owen et al. (2004)	13 (U17)	25 × 30	1 × 3/12 min rest	154 bpm	–	–
Owen et al. (2004)	13 (U17)	30 × 35	1 × 3/12 min rest	163 bpm	–	–
Owen et al. (2004)	13 (U17)	35 × 40	1 × 3/12 min rest	164 bpm	–	–
Williams and Owen (2007)	9 (U17)	30 × 25	–	152 bpm	–	–
Rampinini et al. (2007)[a]	20 (Amateurs)	28 × 20	3 × 4 min/3 min rest	87.8 % HRmax	5.2	7.2 [0–10 scale]
Rampinini et al. (2007)[a]	20 (Amateurs)	35 × 25	3 × 4 min/3 min rest	88.8 % HRmax	5.0	7.6 [0–10 scale]
Rampinini et al. (2007)[a]	20 (Amateurs)	42 × 30	3 × 4 min/3 min rest	88.8 % HRmax	5.8	7.5 [0–10 scale]
Little and Williams (2007)	28 (elite)	45 × 30	4 × 6 min/1 min, 30 s rest	88.7 % HRmax	–	14.4 [0–20 scale]
Kelly and Drust (2009)	8 (elite)	30 × 20	4 × 4 min/2 min rest	91.0 % HRmax	–	–
Kelly and Drust (2009)	8 (elite)	40 × 30	4 × 4 min/2 min rest	90.0 % HRmax	–	–
Kelly and Drust (2009)	8 (elite)	50 × 40	4 × 4 min/2 min rest	89.0 % HRmax	–	–
Da Silva et al. (2011)	17 (U15)	30 × 30	3 × 4 min/3 min rest	86.9 % HRmax	–	–
Castellano et al. (2013)	14 (semi-professional)	55 × 38	3 × 5 min/5 min rest	92.7 % HRmax	–	–
Aguiar et al. (2013)	10 (U18)	150 m^2 per player	3 × 6 min/1 min rest	84.56 % HRmax	–	–

SF Size of the field (m); *HR* Heart rate; *BLa*$^{-1}$ Blood lactate concentration (mmol/L); *RPE* Rated of perceived exertion
[a]HR values in 5 versus 5 with verbal encouragement during task

The time–motion analysis carried out during 5 versus 5 format revealed that 100–110 m per min are covered per players (see Table 4.13). The high-intensity running or sprint represents 12–18 % of the distance covered. These values are very

Table 4.13 Time–motion analysis during 5 versus 5 format

Study	Participants	SF	Regimen	TD	TD 0–6.9	TD 7.0–12.9	TD 13.0–17.9	TD > 18
Castellano et al. (2013)	14 (semi-professional)	55 × 38	3 × 5 min/5 min rest	492.8	173	167	65	24
Aguiar et al. (2013)	10 (U18)	150 m² per player	3 × 6 min/1 min rest	659.98	285.30	260.19	92.49	21.99
Owen et al. (2014)	10 (elite)	46 × 40	3 × 5 min/3 min rest	1552	650	711	185	6

TD Total distance (m); *TD* 0–6.9 Total distance at 0–6.9 km h⁻¹; *TD* 7.0–12.9 Total distance at 7.0–12.9 km h⁻¹; *TD* 13.0–17.9 Total distance at 13.0–17.9 km h⁻¹; *TD* > 18 Total distance at > 18 km h⁻¹

Table 4.14 Technical performance during 5 versus 5 format

Study	Participants	SF	Regimen	Indicator	Indicator	Indicator
Owen et al. (2004)	13 (U17)	25 × 30 30 × 35 35 × 40	1 × 3/12 min rest	6 passes per player	4 receives per player	1 dribble per player
Kelly and Drust (2009)	8 (elite)	30 × 20 40 × 30 50 × 40	4 × 4 min/2 min rest	21 passes	42.25 receives	15 dribbles
Owen et al. (2014)	10 (elite)	30 × 25	3 × 5 min/3 min rest	170.5 passes	129 receives	23 dribbles

similar with the format 4 versus 4 but smaller than the 35 % verified during 3 versus 3 format.

Technical analysis carried out in 5 versus 5 format (Table 4.14) revealed that 5–11 passes per min are made during this format and 2–4 dribbles are made per minute, thus being smaller values than in 4 versus 4 format. The increase of complexity may turn the drill more tactical and with more time required to make the decision, thus being one reason for the small number of passes made.

4.7 6 Versus 6–10 Versus 10 Formats

This section compiled the analyses carried out in medium to large-sided games (see Table 4.15). These games are not so common and for that reason this structure makes easier to compare all of them. Intensities between 81 and 94 % of HRmax and blood lactate concentrations of 4.5–5.0 mmol/L were found during these games. The prescription may vary for each kind of format; nevertheless these larger formats may fit to develop long intensive endurance. For that reason, 3–4 bouts of 4–8 min with 1 min and 30 s–3 min of rest may be adequate to prescribe these games (Clemente et al. 2014a; Little 2009).

The study carried out by Owen et al. (2014) revealed that larger formats increases the distance covered by the players (see Table 4.16). This evidence was also found in the smaller formats. The intensity of running also increases in larger formats, maybe to perform longer distances in sprint to create longer lines of pass and exploit the length of the field.

Table 4.17 shows the technical performance during different large-sided games. A decrease in the number of passes, receives, and dribbles can be seen with the increase in the number of players per format. Moreover, it is hypothesized that the increase of players per format also decreases the number of individual skills performed by each player. For that reason, large-sided games are better to improve collective organization and not recommended for technical development or individual participation.

Table 4.15 Acute physiological effects during 6 versus 6 to 10 versus 10 formats

Study	Format	SF	Regimen	HR	BLa^{-1}	RPE
Rampinini et al. (2007)[a]	6 versus 6	24 × 32	3 × 4 min/3 min rest	86.4 % HRmax	4.5	6.8 [0– 10 scale]
Rampinini et al. (2007)[a]	6 versus 6	30 × 40	3 × 4 min/3 min rest	87.0 % HRmax	5.0	7.3 [0– 10 scale]
Rampinini et al. (2007)[a]	6 versus 6	36 × 48	3 × 4 min/3 min rest	86.9 % HRmax	4.8	7.2 [0– 10 scale]
Little and Williams (2007)	6 versus 6	50 × 30	3 × 8 min/1 min, 30 s rest	87.6 % HRmax	–	13.7 [0– 20 scale]
Katis and Kellis (2009)	6 versus 6	30 × 40	10 × 4 min/3 min rest	82.8 % HRmax	–	–
Castellano et al. (2013)	7 versus 7	64 × 46	3 × 7 min/3 min rest	94.3 % HRmax	–	–
Jones and Drust (2007)	8 versus 8	60 × 40	10 min	168 bpm	–	–
Little and Williams (2007)	8 versus 8	70 × 45	4 × 8 min/1 min, 30 s rest	88.4 % HRmax	–	14.0 [0– 20 scale]
Owen et al. (2011)	9 versus 9	60 × 50	3 × 5 min/4 min rest	81 % HRmax	–	–

SF Size of the field (m); *HR* Heart rate; *BLa*$^{-1}$ Blood lactate concentration (mmol/L); *RPE* Rated of perceived exertion
[a]HR values in 6 versus 6 with verbal encouragement during task

4.8   Summarizing the Differences

This chapter aimed to show the acute responses that the formats of the game induce in soccer players. Greater intensities were generally found in smaller formats (extreme SSCGs—1 vs. 1–3 vs. 3). These games are recommended for anaerobic workout and for that reason duration of 1–3 min is recommended with work-to-rest ratio of 1:1. The intensities are progressively decreasing from 4 versus 4–6 versus 6 the intensities, thus being better formats to short intensive aerobic training, with short periods of time (3–5 min) and a work-to-rest ratio of 1:0.5. Finally, large-sided games (7 vs. 7–10 vs. 10) are recommended for long intensive aerobic training, thus longer periods (4–8 min) can be recommended with 1–3 min of rest between bouts. The following Table 4.18 represents the summary of the differences between formats for the studies that compared different formats. More symbols of (+) indicate greater intensities in heart rate responses.

The time–motion analysis carried out by different studies and showed during this chapter revealed that smaller formats lead to more intensity of running (high-intensity running and sprinting) but with fewer distance covered by players. For that reason, smaller formats are better to increase the intensity and also to workout acceleration and deceleration, thus being also possible to develop power of

Table 4.16 Time–motion analysis during 6 versus 6–10 versus 10 formats

Study	Format	SF	Regimen	TD	TD 0–6.9	TD 7.0–12.9	TD 13.0–17.9	TD > 18
Owen et al. (2014)	6 versus 6	50 × 44	3 × 5 min/3 min rest	1570	620	753	190	8
Owen et al. (2014)	7 versus 7	54 × 45	3 × 5 min/3 min rest	2054	738	1012	281	23
Castellano et al. (2013)	7 versus 7	64 × 46	3 × 7 min/3 min rest	499.1	165	208	89	37
Jones and Drust (2007)	8 versus 8	60 × 40	10 min	693	–	–	–	–
Owen et al. (2014)	8 versus 8	60 × 50	3 × 5 min/3 min rest	1606	618	805	168	16
Owen et al. (2014)	9 versus 9	70 × 56	3 × 5 min/3 min rest	1847	562	909	341	35
Owen et al. (2014)	10 versus 10	80 × 70	3 × 5 min/3 min rest	1750	599	836	254	61

TD Total distance (m); *TD* 0–6.9 Total distance at 0–6.9 km h^{-1}; *TD* 7.0–12.9 Total distance at 7.0–12.9 km h^{-1}; *TD* 13.0–17.9 Total distance at 13.0–17.9 km h^{-1}; *TD* > 18 Total distance at > 18 km h^{-1}

Table 4.17 Technical performance during 6 versus 6–10 versus 10 formats

Study	Participants	SF	Regimen	Indicator	Indicator	Indicator
Katis and Kellis (2009)	6 versus 6	30 × 40	10 × 4 min/3 min rest	35 short passes	11 long passes	5 dribbles
Owen et al. (2014)	6 versus 6	50 × 44	3 × 5 min/3 min rest	170 passes	138.5 receives	22.5 dribbles
Owen et al. (2014)	7 versus 7	54 × 45	3 × 5 min/3 min rest	146 passes	114.5 receives	10.5 dribbles
Owen et al. (2014)	8 versus 8	60 × 50	3 × 5 min/3 min rest	126.5 passes	98.5 receives	10.0 dribbles
Jones and Drust (2007)	8 versus 8	60 × 40	10 min	Individual 13 ball contacts	–	–
Owen et al. (2011)	9 versus 9	60 × 50	3 × 5 min/4 min rest	283 passes	267 receives	11 dribbles
Owen et al. (2014)	9 versus 9	70 × 56	3 × 5 min/3 min rest	115.5 passes	92.5 receives	13.0 dribbles
Owen et al. (2014)	10 versus 10	80 × 70	3 × 5 min/3 min rest	122.5 passes	95.5 receives	18.0 dribbles

Table 4.18 Summary table of the heart rate responses in studies that compare different formats

	1 versus 1	2 versus 2	3 versus 3	4 versus 4	5 versus 5	6 versus 6	7 versus 7	8 versus 8	9 versus 9	10 versus 10
Aroso et al. (2004)		++	+++	+						
Owen et al. (2004)	+++++	+++	++++	+	++					
Little and Williams (2007)		++	++++	+++	++	+		+		
Rampinini et al. (2007)			++++	+++	++	+				
Williams and Owen (2007)	+++++	++++	+++	++	+					
Sampaio et al. (2007)		++	+							
Katis and Kellis (2009)			++							
Hill-Haas et al. (2009)		+++	++	+		+				
Owen et al. (2011)			++						+	
Dellal et al. (2011a, b)		+++	++	+						
Köklü et al. (2011)			+++	++						
Da Silva et al. (2011)		+	++	++	+					
Brandes et al. (2012)		+++	++	+						
Castellano et al. (2013)			++		+		+++			
Aguiar et al. (2013)		+++	++++	++	+					
Clemente et al. (2014a)		++	+	+++						

Legend More symbols of (+) identify the greatest HR responses

lowers limbs during these tasks. On the other hand, larger formats are better to run longer distances and also to keep speed after a short acceleration.

In the case of technical analysis, studies suggest that smaller formats increase the individual actions per player. Moreover, smaller formats also increase the duels and the dribble, thus being recommended to develop both the skills. On other hand, larger formats are better to develop pass and large-sided games are also recommended to increase the longer passes that can be useful to adequate to some tactical principles of coaches. Nevertheless, in novices or youth players, smaller formats can be better to increase the individual participation. On the other hand, larger formats can be more adequate to develop the tactical behavior and the decision-making.

About the tactical topic, a study that used a 5 versus 5 format introduce two tactical metrics: (i) centroid; and (ii) surface area (Frencken et al. 2011). The centroid can be understood as the geometric mean point of all positions of a team. The surface area can be described as the area covered by a polygon constituted by all players. In 10 of 19 goals analyzed, the centroid of the attacking team overtakes the centroid of the defending team, thus unbalanced defenses can justify the majority of the goals scored during SSCGs (Frencken et al. 2011). Moreover, it was also found a synchronization tendency between centroids of teams. Following the use of centroid metric in SSCGs, a study compared 2 versus 2, 3 versus 3, 4 versus 4, and 5 versus 5 formats (Aguiar et al. 2015). In this study, it was found that the distance between centroids presented a small decrease from 2 versus 2–4 versus 4 format and a moderate to nearly perfect increase to 5 versus 5 format (Aguiar et al. 2015). The authors suggested that the absolute distance from the players to both their own team and the opponents' team centroid increased from 2 versus 2 to 5 versus 5 formats, the regularity has also increased across the formats, thus to increase the players' positional regularity it is more recommended larger formats (Aguiar et al. 2015).

References

Aguiar, M. V. D., Botelho, G. M. A., Gonçalves, B. S. V., & Sampaio, J. E. (2013). Physiological responses and activity profiles of football small-sided games. *Journal of Strength and Conditioning Research, 27*(5), 1287–1294.

Aguiar, M., Gonçalves, B., Botelho, G., Lemmink, K., & Sampaio, J. (2015). Footballers' movement behaviour during 2-, 3-, 4- and 5-a-side small-sided games. *Journal of Sports Sciences,* 1–8.

Aroso, J., Rebelo, A. N., & Gomes-Pereira, J. (2004). Physiological impact of selected game-related exercises. *Journal of Sports Sciences, 22*, 522.

Brandes, M., Heitmann, A., & Müller, L. (2012). Physical responses of different small-sided game formats in elite youth soccer players. *Journal of Strength and Conditioning Research, 26*(5), 1353–1360.

Castellano, J., Casamichana, D., & Dellal, A. (2013). Influence of game format and number of players on heart rate responses and physical demands in small-sided soccer games. *Journal of Strength and Conditioning Research, 27*, 1295–1303.

Clemente, F. M., Lourenço, F. M., & Mendes, R. S. (2014a). Developing aerobic and anaerobic fitness using small-sided soccer games: methodological proposals. *Strength and Conditioning Journal, 36*(3), 76–87.

Clemente, F. M., Wong, D. P., Martins, F. M. L., & Mendes, R. S. (2014b). Acute effects of the number of players and scoring method on physiological, physical, and technical performance in small-sided soccer games. *Research in Sports Medicine (Print), 22*(4), 380–397.

Da Silva, C. D., Impellizzeri, F. M., Natali, A. J., de Lima, J. R., Bara-Filho, M. G., Silami-Garçia, E., & Marins, J. C. (2011). Exercise intensity and technical demands of small-sided games in young brazilian soccer players: effect of number of players, maturation, and reliability. *Journal of Strength and Conditioning Research, 25*(10), 2746–2751.

Dellal, A., Chamari, K., Pintus, A., Girard, O., Cotte, T., & Keller, D. (2008). Heart rate responses during small-sided games and short intermittent running training in elite soccer players: A comparative study. *Journal of Strength and Conditioning Research, 22*(5), 1449–1457.

Dellal, A., Chamari, K., Owen, A. L., Wong, D. P., Lago-Penas, C., & Hill-Haas, S. (2011a). Influence of technical instructions on the physiological and physical demands of small-sided soccer games. *European Journal of Sport Science, 11*(5), 341–346.

Dellal, A., Hill-Haas, S., Lago-Penas, C., & Chamari, K. (2011b). Small-sided games in soccer: amateur vs. professional players' physiological responses, physical, and technical activities. *The Journal of Strength and Conditioning Research, 25*(9), 2371–2381.

Frencken, W., Lemmink, K., Delleman, N., & Visscher, C. (2011). Oscillations of centroid position and surface area of football teams in small-sided games. *European Journal of Sport Science, 11*(4), 215–223.

Hill-Haas, S. V., Dawson, B. T., Coutts, A. J., & Rowsell, G. J. (2009). Physiological responses and time–motion characteristics of various small-sided soccer games in youth players. *Journal of Sports Sciences, 27*(1), 1–8.

Hill-Haas, S. V., Dawson, B., Impellizzeri, F. M., & Coutts, A. J. (2011). Physiology of small-sided games training in football. *Sports Medicine, 41*(3), 199–220.

Jones, S., & Drust, B. (2007). Physiological and technical demands of 4 v 4 and 8 v 8 games in elite youth soccer players. *Kinesiology, 39*, 150–156.

Katis, A., & Kellis, E. (2009). Effects of small-sided games on physical conditioning and performance in young soccer players. *Journal of Sports Science and Medicine, 8*(3), 374.

Kelly, D. M., & Drust, B. (2009). The effect of pitch dimensions on heart rate responses and technical demands of small-sided soccer games in elite players. *Journal of Science and Medicine in Sport, 12*(4), 475–479.

Köklü, Y., Asçi, A., Koçak, F. Ü., Alemdaroglu, U., & Dündar, U. (2011). Comparison of the physiological responses to different small-sided games in elite young soccer players. *The Journal of Strength & Conditioning Research, 25*(6), 1522–1528.

Little, T. (2009). Optimizing the use of soccer drills for physiological development. *Strength and Conditioning Journal, 31*(3), 67–74.

Little, T., & Williams, A. G. (2007). Measures of exercise intensity during soccer training drills with professional soccer players. *Journal of Strength and Conditioning Research, 21*, 367–371.

Owen, A., Twist, C., & Ford, P. (2004). Small-sided games: the physiological and technical effect of altering field size and player numbers. *Insight, 7*, 50–53.

Owen, A. L., Wong, D. P., McKenna, M., & Dellal, A. (2011). Heart rate responses and technical comparison between small- vs. large-sided games in elite professional soccer. *Journal of Strength and Conditioning Research, 25*(8), 2104–2110.

Owen, A. L., Wong, D. P., Paul, D., & Dellal, A. (2014). Physical and technical comparisons between various-sided games within professional soccer. *International Journal of Sports Medicine, 35*(4), 286–292.

Rampinini, E., Impellizzeri, F. M., Castagna, C., Abt, G., Chamari, K., Sassi, A., & Marcora, S. M. (2007). Factors influencing physiological responses to small-sided soccer games. *Journal of Sports Sciences, 25*(6), 659–666.

Sampaio, J., Garcia, Macas, V., Ibanez, S., Abrantes, C., & Caixinha, P. (2007). Heart rate and perceptual responses to 2x2 and 3x3 small-sided youth soccer games. *Journal of Sports Science & Medicine, 6*(10), 121–122.

Williams, K., & Owen, A. (2007). The impact of player numbers on the physiological responses to small sided games. *Journal of Sports Science & Medicine, Suppl, 10*, 99–102.

Chapter 5
Acute Effects of Different Sizes of the Field

Abstract The size of the field is one of the main variables that have been analyzed in the aim of the small-sided and conditioned games. Different sizes for the same format of the game influence the acute physiological responses and the time–motion profile of players. For that reason, it is extremely important to identify the most common sizes analyzed in the studies and provide the information to the coaches to optimize the possibilities, and adjust the size of the field in the training context. Therefore, the internal and external training load and the information about the implications in technical actions and tactical behavior will be summarized in the chapter.

Keywords Training load · Size of the field · Small-sided and conditioned games · SSG · Drill-based exercises · Soccer · Football · Sports training

5.1 Introduction

The format of the game influences the acute responses of soccer players as verified in the previous chapter. Nevertheless, the size of the field may also contribute to constrain the activities made in small-sided and conditioned games (SSCG) and for that reason influences the physiological responses and also the technical performance (Clemente et al. 2014). The larger or smaller size of the field will determine the space of play to run or to make decisions. Based on the area of play, the individual space for each player will also be determined. This individual play area of SSCGs can be calculated by dividing the field size by the number of players (Casamichana and Castellano 2010; Fradua et al. 2013).

This chapter will summarize the studies that analyzed the acute effects of different playing areas per player. To make the presentation easier, the tables will be presented per format. Based on this structure, it will be possible to easily verify the most common areas per format and also the length to width ratio that coaches may use to design their SSCGs. A conclusion with some highlights and recommendations will be presented in the end of this chapter.

5.2 Size of the Field: Review of Acute Effects

The size of the field influences the time–motion profile of players. More or less area per player constrains the motion, the actions, and the time to make decisions. The size of the field must consider the area that provides an average per player and for that reason the size will depend from the format of the game (number of players in the task). An analysis to different areas per format will be made in this section.

5.2.1 Comparison of Different Area in 1 Versus 1 Format

Only one study (Owen et al. 2004) compared different sizes of the field in 1 versus 1 format, as far as we know. The heart rate analysis revealed that the biggest areas (75 and 150 m^2) increased the beats per minute (bpm) in comparison with the smaller area (25 m^2). A difference of 6 bpm was identified between the smaller and the bigger formats (see Table 5.1).

5.2.2 Comparison of Different Area in 2 Versus 2 Format

Following the study conducted in 1 versus 1, the same authors (Owen et al. 2004; Williams and Owen 2007) compared the effects of different sizes of the field on 2 versus 2 format (see Table 5.2). The results also revealed that two biggest areas per player (75 and 125 m^2) resulted in an increase of heart rate responses. A difference of 8 bpm was found between the smaller (38 m^2) and the bigger format.

5.2.3 Comparison of Different Area in 3 Versus 3 Format

Four studies (Köklü et al. 2013; Owen et al. 2004; Rampinini et al. 2007; Williams and Owen 2007) that analyzed 3 versus 3 in different field sizes are unanimous in concluding that great sizes increase the acute physiological responses (heart rate,

Table 5.1 Acute physiological effects during 1 versus 1 with different field sizes

Study	Participants	Regimen	SF	APP (m^2)	HR (bpm)	BLa^{-1}	RPE
Owen et al. (2004)	13 (U17)	1 × 3 min/12 min rest	10 × 5	25	176	–	–
			15 × 10	75	181	–	–
			20 × 15	150	182	–	–

SF Size of the field (m); *APP* Area per player; *HR* Heart rate; *BLa*$^{-1}$ Blood lactate concentration (mmol/L); *RPE* Rated of perceived exertion

Table 5.2   Acute physiological effects during 2 versus 2 with different field sizes

Study	Participants	Regimen	SF	APP (m²)	HR (bpm)	BLa⁻¹	RPE
Owen et al. (2004)	13 (U17)	1 × 3 min/12 min rest	15 × 10	38	172	–	–
			20 × 15	75	179	–	–
			25 × 20	125	180	–	–
Williams and Owen (2007)	9 (U17)	–	20 × 15	75	179	–	–
			25 × 20	125	180	–	–

SF Size of the field (m); *APP* Area per player; *HR* Heart rate; *BLa*⁻¹ Blood lactate concentration (mmol/L); *RPE* Rated of perceived exertion

blood lactate concentrations, and perceived exertion). The smaller formats varied between 40 and 50 m², and bigger formats between 90 and 125 m². The studies that used the percentage of maximal heart rate reported values above 90 % in the bigger format, and in smaller formats between 87 and 89.5 % (Köklü et al. 2013; Rampinini et al. 2007). A difference of 0.5 mmol/L between the smaller and the bigger format was found in the unique study (Rampinini et al. 2007) that tested the blood lactate concentrations. The perceived exertion also confirmed the greater effort made in the bigger field (see Table 5.3).

5.2.4   Comparison of Different Area in 4 Versus 4 Format

The studies conducted in 4 versus 4 format verified once again that bigger fields increases the acute physiological responses (see Table 5.4). In this format, it verified that the values were between 82.7 and 90.7 % of HRmax in the bigger fields (100–188 m² per player). The smaller fields (48–75 m²) revealed heart rate responses between 79.1 and 88.7 % of HRmax. A difference between 7 and 8 mmol/L was found between the smallest and the biggest fields. The studies also reported greater values of perceived exertion in the bigger formats.

5.2.5   Comparison of Different Area in 5 Versus 5 Format

Studies carried out in 5 versus 5 format showed more complete information (with technical and time–motion analysis). A unique study found greater heart rate intensities in smaller field than in bigger (Kelly and Drust 2009). The conclusions of the remaining studies followed the evidences described in the smaller formats: bigger area per player increases the acute physiological responses (see Table 5.5). In this format, the smaller fields varied between 56 and 101 m², and the biggest fields between 126 and 273 m². The first study conducted in this format revealed a difference of 10 bpm between the smaller and the bigger field (Owen et al. 2004).

Table 5.3 Acute physiological effects during 3 versus 3 with different field sizes

Study	Participants	Regimen	SF	APP (m^2)	HR	BLa^{-1}	RPE [0-10 scale]
Owen et al. (2004)	13 (U17)	1 × 3 min/12 min rest	15 × 20	50	167 bpm	–	–
			20 × 25	83	167 bpm	–	–
			25 × 30	125	173 bpm	–	–
Rampinini et al. (2007)[a]	20 (Amateurs)	1 × 4 min/3 min rest	12 × 20	40	89.5 % HRmax	6.0	8.1
			15 × 25	63	90.5 % HRmax	6.3	8.4
			18 × 30	90	90.9 % HRmax	6.5	8.5
Williams and Owen (2007)	9 (U17)	–	20 × 15	50	164 bpm	–	–
			25 × 20	83	166 bpm	–	–
			30 × 25	125	171 bpm	–	–
Köklü et al. (2013)	16 (U15)	4 × 3 min/2 min rest	20 × 15	50	176 bpm 87.1 % HRmax	–	5.2
			25 × 18	75	180.1 bpm 89.0 % HRmax	–	5.6
			30 × 20	100	184.2 bpm 91.0 % HRmax	–	6.1

SF Size of the field (m); *APP* Area per player; *HR* Heart rate; *BLa^{-1}* Blood lactate concentration (mmol/L); *RPE* Rated of perceived exertion
[a]HR values with verbal encouragement during task

Differences between 1 and 1.6 % of HRmax were found in the remaining studies (Aslan 2013; Casamichana and Castellano 2010; Hodgson et al. 2014; Rampinini et al. 2007).

The time–motion analysis may provide the justification for the greatest acute physiological responses in bigger fields (Casamichana and Castellano 2010; Hodgson et al. 2014). The study conducted by Casamichana and Castellano (2010) revealed that players covered more 43.66 % of the distance in the bigger field (273 m^2) than in the smaller (74 m^2). In the same study, it was also found that players covered more 8.66 m/min of the distance in sprint in bigger field than in the smaller. An increase of 26.24 % of the distance covered, and a greater distance covered in sprint (including accelerations) in bigger format was also found by Hodgson et al. (2014) (Table 5.6).

Despite the greater intensities found in bigger fields, the technical analysis that compared different sizes in 5 versus 5 revealed that smaller field increases the technical performance (Aslan 2013; Hodgson et al. 2014). More ball possessions, passes, and successful passes were carried out in smaller fields. By the other hand, more dribbles were made in bigger fields (maybe for the increase on the space to try the dribble and the duel) (Table 5.7).

Table 5.4 Acute physiological effects during 4 versus 4 with different field sizes

Study	Participants	Regimen	SF	APP (m²)	HR	BLa⁻¹	RPE
Aroso et al. (2004)	14 (U16)	1 × 1 min, 30 s/1 min, 30 s rest	30 × 20	75	79.1 % HRmax	2.6	13.3 [0–20 scale]
			50 × 30	188	82.7 % HRmax	3.4	14.7 [0–20 scale]
Owen et al. (2004)	13 (U17)	1 × 3 min/12 min rest	20 × 25	63	147 bpm	–	–
			25 × 30	94	160 bpm	–	–
			30 × 35	131	158 bpm	–	–
Rampinini et al. (2007)[a]	20 (Amateurs)	1 × 4 min/3 min rest	16 × 24	48	88.7 % HRmax	5.3	7.6 [0–10 scale]
			20 × 30	75	89.4 % HRmax	5.5	7.9 [0–10 scale]
			24 × 36	108	89.7 % HRmax	6.0	8.1 [0–10 scale]
Köklü et al. (2013)	16 (U15)	4 × 4 min/2 min rest	20 × 20	50	175.0 bpm 86.5 % HRmax	–	4.4 [0–10 scale]
			30 × 20	75	179.9 bpm 88.9 % HRmax	–	5.0 [0–10 scale]
			32 × 25	100	183.5 bpm 90.7 % HRmax	–	5.3 [0–10 scale]

SF Size of the field (m); *APP* Area per player; *HR* Heart rate; *BLa⁻¹* Blood lactate concentration (mmol/L); *RPE* Rated of perceived exertion

[a]HR values with verbal encouragement during task

Table 5.5 Acute physiological effects during 5 versus 5 with different field sizes

Study	Participants	Regimen	SF	APP (m²)	HR	BLa⁻¹	RPE
Owen et al. (2004)	13 (U17)	1 × 3 min/12 min rest	25 × 30	75	154 bpm	–	–
			30 × 35	105	163 bpm	–	–
			35 × 40	140	164 bpm	–	–
Rampinini et al. (2007)[a]	20 (Amateurs)	1 × 4 min/3 min rest	28 × 20	56	87.8 % HRmax	5.2	7.2 [0–10 scale]
			35 × 25	88	88.8 % HRmax	5.0	7.6 [0–10 scale]
			42 × 30	126	88.8 % HRmax	5.8	7.5 [0–10 scale]
Kelly and Drust (2009)	8 (elite)	4 × 4 min/2 min rest	30 × 20	60	91.0 % HRmax	–	–
			40 × 30	120	90.0 % HRmax	–	–
			50 × 40	200	89.0 % HRmax	–	–
Casamichana and Castellano (2010)	10 (U16)	3 × 8 min/5 min rest	32 × 23	74	93.0 % HRmax	–	6.7 [0–10 scale]
			50 × 35	175	94.6 % HRmax	–	6.7 [0–10 scale]
			62 × 44	273	94.6 % HRmax	–	5.7 [0–10 scale]
Aslan (2013)	10 (Recreational)	1 × 40 min	44 × 23	101	79.4 % HRres	–	1.2 [0–20 scale]
			57 × 30	171	81.7 % HRres	–	1.9 [0–20 scale]
Hodgson et al. (2014)	8 (Amateurs)	4 × 3 min/2 min rest	30 × 20	60	164 bpm 86 % HRmax	–	–
			40 × 30	120	168 bpm 87 % HRmax	–	–
			50 × 40	200	168 bpm 87 % HRmax	–	–

SF Size of the field (m); *APP* Area per player; *HR* Heart rate; *BLa⁻¹* Blood lactate concentration (mmol/L); *RPE* Rated of perceived exertion
[a]HR values with verbal encouragement during task

Table 5.6 Time–motion analysis during 5 versus 5 in different field sizes

Study	Regimen	SF	APP (m^2)	TD	TD 0–6.9	TD 7.0–12.9	TD 13.0–17.9	TD > 18
Casamichana and Castellano (2010)	3 × 8 min/5 min rest	32 × 23	74	695.8	401.7	238.9	50.2	4.9
		50 × 35	175	908.9	390.6	329.3	155.4	28.5
		62 × 44	273	999.6	378.2	366.3	180.9	74.2
Hodgson et al. (2014)	4 × 3 min/2 min rest	30 × 20	60	1532	–	–	–	0
		40 × 30	120	1941	–	–	–	40
		50 × 40	200	1934	–	–	–	77

TD Total distance (m); *TD* 0–6.9 Total distance at 0–6.9 km h^{-1}; *TD* 7.0–12.9 Total distance at 7.0–12.9 km h^{-1}; *TD* 13.0–17.9 Total distance at 13.0–17.9 km h^{-1}; *TD* > 18 Total distance at > 18 km h^{-1}

Table 5.7 Technical performance during 5 versus 5 in different field sizes

Study	Participants	Regimen	SF	APP (m^2)	Indicator	Indicator	Indicator (dribbles)
Aslan (2013)	10 (Recreational)	1 × 40 min	44 × 23	101	47.4 ball possessions	27.2 successful passes	13.9
			57 × 30	171	43.4 ball possessions	24.3 successful passes	15.6
Hodgson et al. (2014)	8 (Amateurs)	4 × 3 min/2 min rest	30 × 20	60	22 passes	5 shots	7
			40 × 30	120	21 passes	5 shots	8
			50 × 40	200	20 passes	3 shots	7

SF Size of the field (m); *APP* Area per player

5.2.6   Comparison of Different Area in 6 Versus 6 to 10 Versus 10 Formats

Different field sizes were compared in 6 versus 6 and 7 versus 7 formats. In the study carried out by Rampinini et al. (2007), in 6 versus 6 format it was possible to identify that the smaller format (64 m^2) had the lowest heart rate responses, blood lactate concentrations, and perceived exertion. Nevertheless, in this study, the greater intensities were found in middle size (100 m^2). The study carried in 7 versus 7 format was possible to verify that bigger format increased the heart rate responses (1.9 % of HR max) and the perceived exertion (Table 5.8).

Similar to the study conducted in 5 versus 5, the technical performance was greater in smaller field. More ball possessions, successful passes, and dribbles were made in smaller field during the 7 versus 7 format (Aslan 2013) (Table 5.9).

Table 5.8 Acute physiological effects during large-sided games with different field sizes

Study	Format	Regimen	SF	APP (m^2)	HR	BLa^{-1}	RPE [0-10 scale]
Rampinini et al. (2007)[a]	6 versus 6	1 × 4 min/3 min rest	24 × 32	64	86.4 % HRmax	4.5	6.8
			30 × 40	100	87.0 % HRmax	5.0	7.3
			36 × 48	144	86.9 % HRmax	4.8	7.2
Aslan (2013)	7 versus 7	1 × 40 min	44 × 23	72	76.8 % HRres	–	0.9
			57 × 30	122	78.7 % HRres	–	1.2

SF Size of the field (m); *APP* Area per player; *HR* Heart rate; *BLa^{-1}* Blood lactate concentration (mmol/L); *RPE* Rated of perceived exertion

[a] HR values with verbal encouragement during task

Table 5.9 Technical performance during large-sided games with different field sizes

Study	Format	Regimen (min)	SF	APP (m^2)	Indicator (ball possessions)	Indicator (successful passes)	Indicator (dribbles)
Aslan (2013)	7 versus 7	1 × 40	44 × 23	72	45.0	21.8	12.1
			57 × 30	122	40.5	20.9	11.3

SF Size of the field (m); *APP* Area per player

5.3   Conclusions

The size of the field influences the performance of players during SSCGs. Bigger sizes increased the heart rate responses, blood lactate concentrations, perceived exertion, distance covered, and the distance covered in sprint in all formats that have been studied. In the other hand, better technical performances were achieved in smaller sizes, thus suggesting that the decrease of the space may increase the opportunity to exploit skills. Larger sizes may be more adequate to increase the physiological and physical demands of the game and the smaller formats to develop the technic. To better identify the meaning of smaller and bigger sizes of the field, Table 5.10 summarizes the dimensions used by different authors per format of the game.

Only two studies analyzed the tactical behavior that emerges from different sizes of the field (Frencken et al. 2013; Vilar et al. 2014). A study analyzed the influence of three dimensions (40×20—80 m^2 per player; 52×26—132.5 m^2 per player; and 28×14—39.2 m^2 per player) during 5 versus 5 game in the shaping opportunities to maintain the ball possession, pass to teammates, and shoot at goal (Vilar et al. 2014). The results of this study revealed that interpersonal distances between players were significantly lower in smaller field and afforded greater opportunities to maintain the ball possession (Vilar et al. 2014). Nevertheless, no statistical differences between field sizes were observed for opportunities to shoot at goal and pass to teammates.

In the other study (Frencken et al. 2013), the collective organization of the teams in four different sizes of the field (30×20—75 m^2 per player; 24×20—60 m^2 per player; 30×16—60 m^2 per player; and 24×16—48 m^2 per player) during 4 versus 4 games was analyzed. The results revealed that reducing the field length causes players to close in on each other longitudinally (Frencken et al. 2013). It was also found that the teams' centroids tend to move more in the same direction longitudinally in smaller fields. The decrease in the width of the field reduced the lateral distances between teammates (Frencken et al. 2013).

Both studies (Frencken et al. 2013; Vilar et al. 2014) suggested that smaller sizes increase the capacity to play with small interpersonal distances and increase the capacity to maintain the possession of the ball. It was also suggested that small

Table 5.10 Field sizes considered small, medium, and large in different formats of the game

Format	Small		Medium		Large	
	Dimensions	APP	Dimensions	APP	Dimensions	APP
1 versus 1	10×5	25	15×10	75	20×15	150
2 versus 2	15×10	38	20×15	75	25×20	125
3 versus 3	20×15	50	25×18	75	25×30	125
4 versus 4	20×25	63	30×20	75	30×35	131
5 versus 5	30×20	60	35×25	88	42×30	126
6 versus 6	32×24	64	40×30	100	48×36	144
7 versus 7	40×25	71	44×23	72	57×30	122

Dimensions length × width (m); *APP* Area per player (m^2)

fields also contribute to ensure synchronization between opponent's centroids, thus being an important indication to improve the capacity to flow based on the opponents' dynamics and ball.

Trying to identify the appropriate sizes to design SSCGs, a pilot study determined the individual playing area of players during full-size matches by dividing the area of the rectangle that includes all outfield players by twenty (Fradua et al. 2013). Six goal-to-goal areas split the field and the individual area per player was determined per positioning of the ball in these areas. A larger area per player was verified in the moments the ball circulated in the area closer to the opponent's goal. On the other hand, the smaller area per player was found in the moments the ball circulated in the middle of the field. This study verified that individual area during matches varies between 78.97 and 93.87 m^2. The authors made the following considerations for designing SSCGs according to the particular phase of play (Fradua et al. 2013):

- Build-up play: 90 m^2 [range 70–110] area per player, with length to width ratio of 1:1
- Transition play: 80 m^2 [range 65–95] are per player, with length to width ratio of 1:1.3
- Finishing phase: 90 m^2 [range 70–110] area per player, with length to width ratio of 1:1

These interesting findings can be useful to coaches during the designing moment of the games. Another important issue that may arrive from the use of different sizes is the application during training sessions. Different sizes lead to different places to organize the task. The use of games requires some visual marks of the boundaries. Nevertheless, coach should save time to organize these fields. For that reason, he may use the boundaries of a full soccer field to reduce some time in place the visual marcs. Let us provide in Fig. 5.1, the official sizes of a soccer field.

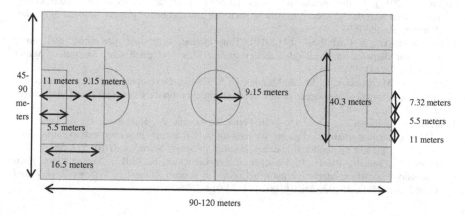

Fig. 5.1 Standard soccer field measurements

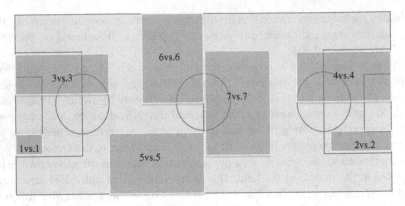

Fig. 5.2 A possible place to develop different formats of the game for smaller sizes

Coaches may use some specific places in the field to mark the zone for SSCGs. We would like to proposesome specific places in the field for some formats Fig. 5.2.

In summary, this chapter found that bigger sizes increases the acute physiological and physical responses, and thus are more appropriate to develop the fitness. On the other hand, smaller sizes are more appropriate for technical performance and to increase the tactical behavior and collective organization of the teams. These findings should be considered in the moment of designing SCCGs to soccer players in different stages and competitive levels.

References

Aroso, J., Rebelo, A. N., & Gomes-Pereira, J. (2004). Physiological impact of selected game-related exercises. *Journal of Sports Sciences, 22*, 522.

Aslan, A. (2013). Cardiovascular responses, perceived exertion and technical actions during small-sided recreational soccer: effects of pitch size and number of players. *Journal of Human Kinetics, 38*, 95–105.

Casamichana, D., & Castellano, J. (2010). Time–motion, heart rate, perceptual and motor behaviour demands in small-sides soccer games: Effects of pitch size. *Journal of Sports Sciences, 28*(14), 1615–1623.

Clemente, F. M., Martins, F. M., & Mendes, R. S. (2014). Developing aerobic and anaerobic fitness using small-sided soccer games: methodological proposals. *Strength and Conditioning Journal, 36*(3), 76–87.

Fradua, L., Zubillaga, A., Caro, O., Iván Fernández-García, A., Ruiz-Ruiz, C., & Tenga, A. (2013). Designing small-sided games for training tactical aspects in soccer: extrapolating pitch sizes from full-size professional matches. *Journal of Sports Sciences, 31*(6), 573–581.

Frencken, W., van der Plaats, J., Visscher, C., & Lemmink, K. (2013). Size matters: pitch dimensions constrain interactive team behaviour in soccer. *Journal of Systems Science and Complexity, 26*(1), 85–93. doi:10.1007/s11424-013-2284-1.

Hodgson, C., Akenhead, R., & Thomas, K. (2014). Time-motion analysis of acceleration demands of 4v4 small-sided soccer games played on different pitch sizes. *Human Movement Science, 33,* 25–32.

Kelly, D. M., & Drust, B. (2009). The effect of pitch dimensions on heart rate responses and technical demands of small-sided soccer games in elite players. *Journal of Science and Medicine in Sport, 12*(4), 475–479.

Köklü, Y., Albayrak, M., Keysan, H., Alemdaroğlu, U., & Dellal, A. (2013). Improvement of the physical conditioning of young soccer players by playing small-sided games on different pitch size - special reference to physiological responses. *Kinesiology, 45*(1), 41–47.

Owen, A., Twist, C., & Ford, P. (2004). Small-sided games: the physiological and technical effect of altering field size and player numbers. *Insight, 7,* 50–53.

Rampinini, E., Impellizzeri, F. M., Castagna, C., Abt, G., Chamari, K., Sassi, A., & Marcora, S. M. (2007). Factors influencing physiological responses to small-sided soccer games. *Journal of Sports Sciences, 25*(6), 659–666.

Vilar, L., Duarte, R., Silva, P., Chow, J. Y., & Davids, K. (2014). The influence of pitch dimensions on performance during small-sided and conditioned soccer games. *Journal of Sports Sciences, 32*(19), 1751–1759.

Williams, K., & Owen, A. (2007). The impact of player numbers on the physiological responses to small sided games. *Journal of Sports Science & Medicine, Suppl, 10,* 99–102.

Chapter 6
Adjusting the Design: New Rules to Maximize the Experience

Abstract The aim of this chapter is to provide the information about the implications of changing the rules of small-sided games. By using the data from the most recent studies, a set of conditioned games will be analyzed in their acute effects, particularly in internal and external training load and in the technical and tactical adjustments that depend on the game. Task conditions such as the type of targets or goals, goalkeepers, limits on ball touches, numerical relationship with the opponent or the type of encouragement will be analyzed, helping the coaches to identify the best variables to constrain in the particular contexts.

Keywords Training load · Task constraints · Small-sided and conditioned games · SSG · Drill-based exercises · Soccer · Football · Sports training

6.1 Introduction

The modifications of the format and size of the field are very common (Aguiar et al. 2012; Halouani et al. 2014b; Hill-Haas et al. 2011). Nevertheless, small-sided and conditioned games (SSCGs) are often associated with changes in the rules and the dynamic of the game (Davids et al. 2013). These changes aim to increase the execution of some skills or augment the perception for specific tactical issues (Renshaw et al. 2015). Nevertheless, there are also implications for acute physiological responses and time–motion profile. For that reason, this chapter will show some typical task constraints (or conditions) used by coaches and identify the main results.

This chapter will be organized by regular conditions used in soccer training: (i) different goals/targets; (ii) rules and objective of the game; (iii) the use of goalkeepers; (iv) training regimen; (v) the use of floaters/neutral players; and (vi) coach encouragement. In the end of the chapter a summary of the findings will be presented.

© The Author(s) 2016
F.M. Clemente, *Small-Sided and Conditioned Games in Soccer Training*,
SpringerBriefs in Applied Sciences and Technology,
DOI 10.1007/978-981-10-0880-1_6

6.2 Changing the Goals

The use of different goals influence the tactical issue in the game and also the type of movements carried out by players. For that reason, this section will show the research carried out about this task condition (Table 6.1). Usually, SSCGs are played without official goals. For that reason, some coaches use small goals and others opt to use lines of goal (to pass through in dribble or to stop the ball). Both conditions influence the acute physiological responses differently.

The first study that compared different games with and without goals were carried out in a 3 versus 3 format (Mallo and Navarro 2008). The authors tested

Table 6.1 Acute physiological effects in different conditions with and without goals

Study	F/SF	Regimen	Condition	HR	BLa^{-1}	RPE [0–10 scale]
Mallo and Navarro (2008)	3 versus 3 33 × 20	3 × 5 min/10 min recovery	Ball possession	173 bpm	–	–
			Jokers (out players)	173 bpm	–	–
			Goals with GK	166 bpm	–	–
Casamichana et al. (2011)	4 versus 4 25 × 32	3 × 4 min/3 min recovery	No goals–ball possession	166.3 bpm	–	–
			Small goals	162.9 bpm	–	–
			Regular goals with GK	161.4 bpm	–	–
Casamichana et al. (2012)	3 versus 3 43 × 30	3 × 6 min/5 min rest	No goals–ball possession	–	–	4.9
			Small goals	–	–	4.2
			Regular goals with GK	–	–	4.1
	5 versus 5 55 × 38		No goals–ball possession	–	–	4.7
			Small goals	–	–	3.5
			Regular goals with GK	–	–	3.3
	7 versus 7 64 × 46		No goals–ball possession	–	–	4.0
			Small goals	–	–	3.6
			Regular goals with GK	–	–	3.0
Halouani et al. (2014a)	3 versus 3 20 × 15	4 × 4 min/2 min recovery	Stop the ball in the line	178 bpm	4.66	7
			Small goals	174 bpm	4.16	6.58

(continued)

Table 6.1 (continued)

Study	F/SF	Regimen	Condition	HR	BLa⁻¹	RPE [0–10 scale]
Clemente et al. (2014b)	2 versus 2 + 2 floaters 19 × 19	3 × 5 min/3 min rest	Line goal	74.98 % HRres	–	–
			Two small goals	81.05 % HRres	–	–
			One small and central goal	83.38 % HRres	–	–
	3 versus 3 + 2 floaters 23 × 23		Line goal	82.06 % HRres	–	–
			Two small goals	84.18 % HRres	–	–
			One small and central goal	81.98 % HRres	–	–
	4 versus 4 + 2 floaters 27 × 27		Line goal	81.27 % HRres	–	–
			Two small goals	80.32 % HRres	–	–
			One small and central goal	83.61 % HRres	–	–

F Format; *SF* Size of the field (m); *HR* Heart rate; *BLa*⁻¹ Blood lactate concentration (mmol/L); *RPE* Rated of perceived exertion; *GK* Goalkeeper

three conditions (see Fig. 6.1): (i) ball possession without goals; (ii) ball possession with two 'jokers' that are outer players that can pass the ball to a player from the team that it was received from; and (iii) game with goals and goalkeepers. The results showed that total distance covered in game with goals was statistically lower and with lower running intensity than the other conditions. Moreover, the game of ball possession without 'jokers' increased the heart rate responses and the time spent in higher effort (Mallo and Navarro 2008).

A study that used 4 versus 4 format (25 × 15 meters) with a regimen of 3 × 15 min/6 min recovery tested the influence of three scoring methods (see Fig. 6.2) on the heart rate fluctuations (Duarte et al. 2010). Results revealed that line goal increased the randomness of heart rate responses. Moreover, line goal also contributes for a more standardized cardiovascular stimulation of the players involved (Duarte et al. 2010).

Differences in heart rate responses between no goal, small goals, and official goals with goalkeepers were found in 4 versus 4 format (Casamichana et al. 2011). Generally, games without goalkeeper statistically increased the heart rate responses. Results found statistical great heart rate responses between central players (defenders and forwards) and external players (defenders and midfielders) in the game

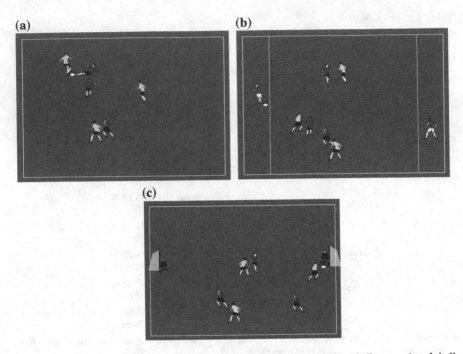

Fig. 6.1 SSCGs with different goals used by Mallo and Navarro (2008): **a** ball possession; **b** ball possession with 'jokers'; and **c** goals with goalkeepers

with small goals. It was also found statistical lower heart rate responses of central players with midfielders in the game are without goals. Finally, central players had statistically lower heart rate responses that the remaining positions in the game with official goals and goalkeepers (Casamichana et al. 2011).

Using 3 formats (3 vs. 3, 5 vs. 5, and 7 vs. 7) it was compared the non-use and the use of small and official goals on the perceived exertion of soccer players (Casamichana et al. 2012). This study confirmed a statistical increase of internal load in tasks without goals, thus following the previous work of the authors in 4 versus 4 format (Casamichana et al. 2011).

The physiological effects of stoping the ball in a line and scoring in small goals were compared in 3 versus 3 format (Halouani et al. 2014b). The results showed that stoping the ball in the line induced statistical greater heart rate responses and blood lactate concentrations than the game with small goals (Halouani et al. 2014b).

Testing the 2 versus 2, 3 versus 3, and 4 versus 4 format in three conditions (line goal, two small goals, and one goal) it was verified that in 2 versus 2 and 3 versus 3 formats players covered greater distances in the game with line goal (Clemente et al. 2014b). It was also found greater heart rate responses in the game with double goals played at 2 versus 2 and 4 versus 4 formats.

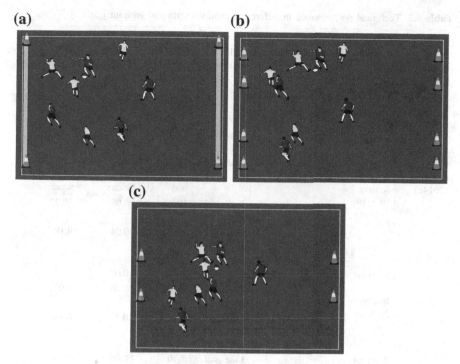

Fig. 6.2 SSCGs with different goals used by Duarte et al. (2010): **a** line goal, **b** double goal; and **c** central goal

The technical performance was only analyzed by two studies (Clemente et al. 2014b; Mallo and Navarro 2008). The study conducted by Mallo and Navarro (2008) revealed that the condition of ball possession without goals increased the contacts with ball and the passes made. On the other hand, the game with goals and goalkeepers resulted in a lower number of passes made.

The study conducted by Clemente et al. (2014b) was inconclusive about the technical effects of different tasks. No statistical evidences were found and the performance varied between the three formats tested (see Table 6.2).

In summary, this section showed that in the majority of the studies it was found the evidence that games without goals increases the heart rate responses, perceives exertion and blood lactate concentration. It was also possible to verify that small goals increase the acute physiological responses in comparison with official goals with goalkeepers. In the case of technical performance the results are inconclusive, nevertheless, there is a small tendency to perform greater volume of skills in conditions without goals.

Table 6.2 Technical performance in different conditions with and without goals

Study	F/SF	Regimen	Condition	Indicator	Indicator	Indicator
Mallo and Navarro (2008)	3 versus 3 33 × 20	3 × 5 min/10 min recovery	Ball possession	20.8 contacts with the ball	16.8 short passes	15.5 % wrong passes
			Jokers (out players)	11.3	10.0	5.3 %
			Goals with GK	12.3	7.5	9.1 %
Clemente et al. (2014b)	2 versus 2 + 2 floaters 19 × 19	3 × 5 min/3 min rest	Line goal	18.33 volume of play	0.22 efficiency index	11.34 performance score
			Two small goals	19.83	0.08	10.72
			One small and central goal	17.50	0.04	9.18
	3 versus 3 + 2 floaters 23 × 23		Line goal	11.88	0.00	5.94
			Two small goals	11.50	0.03	6.04
			One small and central goal	12.25	0.30	6.42
	4 versus 4 + 2 floaters 27 × 27		Line goal	7.70	0.09	4.76
			Two small goals	7.00	0.08	4.30
			One small and central goal	7.30	0.03	3.93

F Format; *SF* Size of the field (m); *GK* Goalkeeper

6.3   Conditioning the Ball Touches

Coaches may limit the number of touches that each player may consecutively perform to increase the speed of decision making and technical performance. Such condition may also influence the intensity of the game and for that reason some studies have been analyzed this task constraint (Aroso et al. 2004; Casamichana et al. 2014; Dellal et al. 2011; Román-Quintana et al. 2013). The main physiological responses found by the studies may be observed in Table 6.3. The study conducted by Aroso et al. (2004) compared free play with three touches limitation in 3 versus 3 format. Tree touches statistically increased the running activity and statistically reduced the time spent in walk and sprint (Aroso et al. 2004). Moreover, three touches had smaller values of heart rate and greater values of blood lactate concentrations and perceived exertion.

Table 6.3 Acute physiological effects in different conditions of ball touches

Study	F/SF	Regimen	Condition	HR	BLa^{-1}	RPE
Aroso et al. (2004)	3 versus 3 30 × 20	3 × 4 min/1 min, 30 s rest	Free Play	81.1 % HRres	4.9	14.5 [0–20 scale]
			Three touches	79.3 % HRres	5.3	15.4 [0–20 scale]
Dellal et al. (2011)	2 versus 2 20 × 15	4 × 2 min/3 min rest	One touch	90.3 % HRmax	3.9	8.2 [0–10 scale]
			Two touches	90.1 % HRmax	3.5	7.7 [0–10 scale]
			Free play	90.0 % HRmax	3.4	7.6 [0–10 scale]
	3 versus 3 25 × 18	4 × 3 min/3 min rest	One touch	90.0 % HRmax	3.8	8.1 [0–10 scale]
			Two touches	89.3 % HRmax	3.3	7.9 [0–10 scale]
			Free play	89.6 % HRmax	3.0	7.5 [0–10 scale]
	4 versus 4 30 × 20	4 × 4 min/3 min rest	One touch	87.6 % HRmax	2.9	8.0 [0–10 scale]
			Two touches	85.6 % HRmax	2.8	7.9 [0–10 scale]
			Free play	84.7 % HRmax	2.9	7.2 [0–10 scale]
Casamichana et al. (2013b)	6 versus 6 + 2 40 × 28	12 min	One touch	93.2 % HRmax	–	–
			Two touches	92.6 % HRmax	–	–
			Free play	90.4 % HRmax	–	–
Román-Quintana et al. (2013)	7 versus 7 with GK 60 × 49	12 min	One touch	89.2 % HRmax	–	–
			Two touches	92.7 % HRmax	–	–
			Free play	90.2 % HRmax	–	–

(continued)

Table 6.3 (continued)

Study	F/SF	Regimen	Condition	HR	BLa^{-1}	RPE
Casamichana et al. (2014)	6 versus 6 60 × 49	2 × 6 min	Two touches	83.8 % HRmax (1st bout) 89.3 % HRmax (2nd bout)	–	–
			Free play	89.0 % HRmax (1st bout) 90.4 % HRmax (2nd bout)	–	–

F Format; *SF* Size of the field (m); *HR* Heart rate; *BLa*$^{-1}$ Blood lactate concentration (mmol/L); *RPE* Rated of perceived exertion; *GK* Goalkeeper

The heart rate responses were also analyzed in 2 versus 2, 3 versus 3, and 4 versus 4 formats in the conditions of one touch, two touches, and free play (Dellal et al. 2011). The main results revealed that one touch had the greatest heart rate values in 2 versus 2 and 3 versus 3 formats and the greatest blood lactate concentrations and perceived exertions in all formats. Large-sided games of 6 versus 6 and 7 versus 7 were analyzed by Casamichana et al. (2013b, 2014) and Román-Quintana et al. (2013). The conditions of one touch, two touches and free play were compared. In the case of Casamichana et al. (2013b) greater values of heart rate were found in one touch condition. On the other hand, in the study conducted by Román-Quintana et al. (2013) showed that the greatest values were found in two touches condition. Finally, in a comparison between two touches and free play, Casamichana et al. (2014) revealed greater values of heart rate in free play in 6 versus 6 format.

The distance covered by it was also analyzed in the previous studies that compared free play with limitation in ball touches (see Table 6.4). In the study conducted by Dellal et al. (2011) greater values of distance covered were found in one touch condition played at 2 versus 2, 3 versus 3, and 4 versus 4 formats. Moreover, the authors also found greater values of distance covered in high intensities and sprint. On the other hand, studies conducted in 6 versus 6 and 7 versus 7 formats revealed greater values of distance covered in the condition of free play (Casamichana et al. 2013b, 2014; Román-Quintana et al. 2013).

Only two studies (Almeida et al. 2012; Dellal et al. 2011) analyzed the technical performance during games with touches limitations, as far we know. The summary of the results can be found in Table 6.5. Della et al. (2011) revealed that in 2 versus 2 and 4 versus 4 format, free play increased the number of duels and successful passes. In the other hand, the possessions were greater in one touch condition played at all formats. In the other study, conditions of two touches, free play, and obligation of four consecutive passes before finalization were compared in 3 versus

Table 6.4 Time–motion profile in different conditions of ball touches

Study	Participants	SF	Regimen	TD	TD 0-6.9	TD 7.0-12.9	TD 13.0-17.9	TD > 18
Dellal et al. (2011)	2 versus 2 20 × 15	4 × 2 min/3 min rest	One touch	1305.5	–	–	330.0	232.3
			Two touches	1211.8	–	–	271.3	195.1
			Free play	1157.7	–	–	245.4	177.5
	3 versus 3 25 × 18	4 × 3 min/3 min rest	One touch	2247.6	–	–	523.2	397.0
			Two touches	2124.7	–	–	473.9	351.2
			Free play	2013.9	–	–	422.4	315.6
	4 versus 4 30 × 20	4 × 4 min/3 min rest	One touch	3057.3	–	–	638.9	493.2
			Two touches	2814.6	–	–	562.0	438.0
			Free play	2663.6	–	–	482.7	381.8
Casamichana et al. (2013b)	6 versus 6 6 + 2 40 × 28	12 min	One touch	1295.2	–	–	–	–
			Two touches	1393.9	–	–	–	–
			Free play	1409.7	–	–	–	–
Román-Quintana et al. (2013)	7 versus 7 with GK 60 × 49	12 min	One touch	1226.8	–	–	–	–
			Two touches	1224.9	–	–	–	–
			Free play	1345.2	–	–	–	–
Casamichana et al. (2014)	6 versus 6 60 × 49	2 × 6 min	Two touches	680.7 (1st bout) 683.0 (2nd bout)				
			Free play	716.3 (1st bout) 642.2 (2nd bout)				

F Format; *SF* Size of the field (meters); *TD* Total distance (m); *TD* 0–6.9 Total distance at 0–6.9 km h⁻¹; *TD* 7.0–12.9 Total distance at 7.0–12.9 km h⁻¹; *TD* 13.0–17.9 Total distance at 13.0–17.9 km h⁻¹ *TD* >18 Total distance at > 18 km h⁻¹

Table 6.5 Technical performance in different conditions of ball touches

Study	F/SF	Regimen	Condition	Indicator	Indicator	Indicator
Dellal et al. (2011)	2 versus 2 20 × 15	4 × 2 min/3 min rest	One touch	17.1 duels	42.5 % successful passes	50.6 possessions
			Two touches	28.5	60.5 %	41.4
			Free play	26.1	66.4 %	40.9
	3 versus 3 25 × 18	4 × 3 min/3 min rest	One touch	30.9	52.0 %	51.8
			Two touches	28.1	69.9 %	43.7
			Free play	26.8	71.0 %	41.7
	4 versus 4 30 × 20	4 × 4 min/3 min rest	One touch	18.0	49.8 %	41.6
			Two touches	16.5	68.9 %	34.7
			Free play	25.1	73.4 %	31.5
Almeida et al. (2012)	3 versus 3 with GK 40 × 30	2 × 5 min/1 min rest	Free play	2.52 passes	10.22 ball touches	12.63 s duration of ball possession
			Two touches	2.33	5.77	9.52 s
			Four passes[a]	6.16	17.57	20.21 s

F Format; *SF* Size of the field (m); *GK* Goalkeeper
[a]Teams had to perform at least four consecutive passes to finalize the attack

3 format with goalkeeper (Almeida et al. 2012). Results revealed that the obligation of four passes increased the number of passes, ball touches, and duration of ball possession. On the other hand, the smaller values for these indicators were found in two touches condition.

In summary, the majority of the studies suggest that limitation in ball touches increase the heart rate responses, blood lactate concentrations, and perceived exertion. In smaller formats (2 vs. 2 to 4 vs. 4) the limitations in ball touches also increases the distances covered and the intensity of the running. Nevertheless, greater values of distance covered were found in large-sided games (6 vs. 6 and 7 vs. 7). In the case of technical performance, limitation in ball touches decreased the number of duels and successful passes. For that reason, games with touches limitations can be appropriate to exploit the intensity and dynamic of the game but is not adequate to improve technical skills such as passes or even the decision making in young and novice players.

6.4   Type of Marking

The type of defensive marking may constraint the intensity of exercise and also the technical performance. Based on this concept, four studies (as far we analyzed) researched the influence of type of marking during SSCGs (Aroso et al. 2004; Casamichana et al. 2015; Ngo et al. 2012; Sampaio et al. 2007). The physiological responses found on these studies can be found in Table 6.6.

The first study was conducted by Aroso et al. (2004) and compared the 2 versus 2 format (30 × 20 meters) with and without individual marking (man-to-man marking). The task had 3 bouts of 1 min and 30 s with a work-to-rest ratio of 1:1. The results showed that blood lactate concentrations were greater in individual marking (9.7 mmol/L vs. 8.1 mmol/L without individual marking) and the time spent in walking was smaller. The heart rate was lower in individual marking than in free play (75.8 % HRres and 77.1 % HRres, respectively).

The study carried out by Sampaio et al. (2007) found that perceived exertion was statistically greater in man-to-man marking (in 2 vs. 2 and 3 vs. 3 formats), nevertheless, the heart rate responses were greater in free play of 2 versus 2 format than in man-to-man marking. Ngo et al. (2012) also tested the 3 versus 3 format to compare man-to-man and free play marking. With and without goals constraints were also used. Greater values of heart rate responses and perceived exertion were found in man-to-man marking (Ngo et al. 2012).

The fourth study that compared man-to-man marking and free play was conducted for three different formats (3 vs. 3, 6 vs. 6, and 9 vs. 9) (Casamichana et al. 2015). Heart rate responses were greater in man-to-man marking in 3 versus 3 and 9 versus 9 formats. The time motion analysis revealed that players covered greater distances in man-to-man marking in 3 versus 3 and 6 versus 6 formats. The greater speeds were found in free play in the 6 versus 6 and 9 versus 9 formats.

Cihan (2015) tested three conditions of marking in 3 versus 3 format. The time–motion profile revealed that double man marking condition and man-to-man marking increased the distance covered and the distances covered in high intensity and sprint during the game. Moreover, Cihan (2015) also found that heart rate values, blood lactate concentrations, and perceived exertions were statistically greater than free play.

In summary, the majority of the studies revealed that man-to-man marking increases the acute physiological responses and also the distances covered by players. Players may need to run greater distances to mark the direct opponent and for that reason also increases the physiological load during the exercise. None of the fourth studies analyzed the technical performance.

Table 6.6 Acute physiological effects in man-to-man and free play conditions

Study	F/SF	Regimen	Condition	HR	BLa^{-1}	RPE
Aroso et al. (2004)	2 versus 2 30 × 20	3 × 1 min, 30 s/1 min, 30 s rest	Man-to-man	75.8 % HRres	9.7	16.7 [0–20 scale]
			Free play	77.1 % HRres	8.1	16.2 [0–20 scale]
Sampaio et al. (2007)	2 versus 2 30 × 20	2 × 1 min, 30 s/1 min, 30 s rest	Man-to-man	80.8 % HRmax	–	17.1 [0–20 scale]
			Free play	81.2 % HRmax	–	14.1 [0–20 scale]
	3 versus 3 30 × 20	2 × 3 min/3 min rest	Man-to-man	80.8 % HRmax	–	16.5 [0–20 scale]
			Free play	79.8 % HRmax	–	14.4 [0–20 scale]
Ngo et al. (2012)	3 versus 3 with goal 25 × 18	3 × 4 min/4 min rest	Man-to-man	80.5 % HRres	–	7.1 [0–10 scale]
			Free play	75.7 % HRres	–	6.0 [0–10 scale]
	3 versus 3 no goal 25 × 18		Man-to-man	80.5 % HRres	–	7.4 [0–10 scale]
			Free play	76.1 % HRres	–	6.9 [0–10 scale]

(continued)

Table 6.6 (continued)

Study	F/SF	Regimen	Condition	HR	BLa^{-1}	RPE
Casamichana et al. 2015)	3 versus 3 29 × 19	6 min	Man-to-man	92.6 % HRmax	–	–
			Free play	91.4 % HRmax	–	–
	6 versus 6 40 × 28	6 min	Man-to-man	90.5 % HRmax	–	–
			Free play	93.2 % HRmax	–	–
	9 versus 9 55 × 30	6 min	Man-to-man	89.0 % HRmax	–	–
			Free play	88.9 % HRmax	–	–
Cihan (2015)	3 versus 3 35 × 20	12 min	Man-to-man	84.83 % HRmax	5.75	4.33 [0–10 scale]
			Double man pressure	88.50 % HRmax	7.13	7.16 [0–10 scale]
			Free play	75.00 % HRmax	8.96	2.00 [0–10 scale]

F Format; *SF* Size of the field (m); *HR* Heart rate; *BLa^{-1}* Blood lactate concentration (mmol/L); *RPE* Rated of perceived exertion

6.5 Exploring the Numerical Unbalance and the Floaters

Floaters can be used to provide numerical superiority to a team with or without the possession of the ball. Five studies analyzed the influence of this condition, as far we know (Bekris et al. 2012; Clemente et al. 2015; Evangelos et al. 2012; Hill-Haas et al. 2010; Sampaio et al. 2014).

The study conducted by Hill-Haas et al. (2010) compared the 3 versus 4, 3 versus 3 (+1 floater), 5 versus 6, and 5 versus 5 (+1 floater) formats. Results found that floaters covered greater distances in small format and completed more sprints in large formats. This can be justified by the frequent changes in ball possession, thus increasing the participation and the activity of floaters.

Greater heart rate values in numerical inferiority were possible to verify in a study that compared the numerical superiority and inferiority during 5 versus 5 format (Sampaio et al. 2014). The authors suggested that that a team in inferiority is required to perform additional work in offense and defense. It was also be possible to verify that numerical inferiority may increase collective decisions of reducing the team area of play and also to decide for prestructured strategical behaviors (Sampaio et al. 2014).

Two studies conducted for the same research team analyzed the effects of using floaters to give numerical superiority during attacking and defensive moments. (Bekris et al. 2012; Evangelos et al. 2012). In the 1-a-side and 4-a-side games, the highest heart rates were observed without a floater player. In the 3-a-side game, the highest heart rates were achieved with a defensive neutral, and in the 2-a-side game, the highest heart rates were observed with an offensive floater (Clemente et al. 2014a). No tendencies were possible to be found from both studies.

Finally, a different study analyzed the heart rate responses of floaters in different formats (1 vs. 1, 2 vs. 2, 3 vs. 3, and 4 vs. 4, with more two floaters in the wings of the field) (Clemente et al. 2015). The highest heart rate responses of floaters were found mostly in the biggest format. The heart rate values of floaters varied between 50 and 56 % of HRreserve and for that reason it can be suggested for very light or recovery workout (Clemente et al. 2015).

In summary, the use of numerical inferiority seems to increase the intensity and acute physiological responses during exercise. Floaters in the game may be constrained to run greater distances. Nevertheless, very light values of intensity may be achieved in floaters that only may act outside of boundaries of the game. In this particular case, the use of floaters may be also used to actively recover from an intense workout. In a example, a coach may prescribe 1 versus 1 + 2 floaters with 4 bouts and a work-to-rest ratio of 1:1. In this case, the players in 1 versus 1 just need to change the positions with the floaters and will participate in an activity during the resting time. With this strategy it can be possible to save some time of training and optimize the transitions and pause between tasks.

6.6   With or Without Encouragement

The use of verbal instructions to encourage the players has been also analyzed as a task condition (Rampinini et al. 2007; Sampaio et al. 2007). The study conducted by Rampinini et al. (2007) compared four formats (3 vs. 3, 4 vs. 4, 5 vs. 5, and 6 vs. 6) of the game with and without coaches' encouragement. The results revealed that with encouragement players had statistically greater heart rate responses, blood lactate concentrations and perceived exertion in all formats of the game. Sampaio et al. (2007) also studied the effect of encouragement in the acute physiological responses at 2 versus 2 and 3 versus 3 formats. The results revealed greater heart rate intensities and perceived exertion in the drills with coaches' encouragement. Both studies suggested that coaches' encouragement might improve the commitment and motivation of players during the task and for that reason increase the intensity of exercise.

6.7   Regimen

SSCGs are commonly used as an intermittent exercise (with bouts and work-to-rest ratios) (Clemente et al. 2014b). A few number of studies compared continuous versus intermittent regimens (Fanchini et al. 2011; Hill-Haas et al. 2009; Köklü 2012). The main results can be found in the following Table 6.7.

In the study conducted by Hill-Haas et al. (2009) found greater values of heart rate, blood lactate concentrations, and perceived exertion in the drills with continuous regimens. No statistical differences were found between intermittent and continuous regimen in distance covered or distance traveled while walking, jogging, or running at moderate speed, nevertheless, statistical differences were found in high-intensity running (Hill-Haas et al. 2009).

A similar comparison between intermittent and continuous regimens was tested in 2 versus 2, 3 versus 3, and 4 versus 4 formats (Köklü 2012). No statistical differences were found in the percentage of maximal heart rate, nevertheless, intermittent regimen had greater intensities in 2 versus 2 and 4 versus 4 formats. Nevertheless, blood lactate concentrations were greater in continuous regimen during 4 versus 4 and 6 versus 6 formats.

In a different study, Fanchini et al. (2011) compared the effects of different intermittent durations. Greater intensities were found in longer periods if not considered the first minute of exercise. In the same study, the analysis of technical performance revealed that greater values of passes, unsuccessful passes, and interceptions were made per minute in the regimen of 2 min (Fanchini et al. 2011). Nevertheless, these differences were not statistically different. Finally, in the study of Casamichana et al. (2013b) it was found that total distance covered was greater during intermittent games, specifically at moderate running speed.

Table 6.7 Acute physiological effects and distance covered in different training regimens

Study	F/SF	Regimen	HR (% HRmax)	BLa-1	RPE	DC
Hill-Haas et al. (2009)	2 versus2 28 × 21	4 × 6 min/1 min, 30 s rest	84	4.8	11.6 [0–20 scale]	2621
	4 versus 4 40 × 30 6 versus 6 49 × 37	24 min	87	5.5	12.3 [0–20 scale]	2596
Fanchini et al. (2011)	3 versus 3 37 × 31	3 × 2 min/4 min recovery	82.4	–	6.7 [0–10 scale]	–
		3 × 4 min/4 min recovery	85.9	–	6.8 [0–10 scale]	–
		3 × 6 min/4 min recovery	85.6	–	6.8 [0–10 scale]	–
Köklü (2012)	2 versus 2 15 × 20	3 × 2 min	88.6	7.8	–	–
		6 min	88.8	8.1	–	–
	3 versus 3 18 × 24	3 × 3 min	92.0	6.8	–	–
		9 min	91.2	7.2	–	–
	4 versus 4 24 × 36	3 × 4 min	90.1	6.7	–	–
		12 min	89.3	6.9	–	–
Casamichana et al. (2013a)	5 versus 5 55 × 38	2 × 8 min/2 min rest	87.1	–	–	–
		4 × 4 min/1 min rest	87.5	–	–	–
		16 min	87.5	–	–	–

F Format; *SF* Size of the field (m); *HR* Heart rate; *BLa*$^{-1}$ Blood lactate concentration (mmol/L); *RPE* Rated of perceived exertion; *DC* Distance covered

Without a consensus, the studies suggest that continuous regimens increase the effort and the acute physiological responses. Nevertheless, better technical performance occurs in intermittent regimens. The smaller fatigue during intermittent regimens may justify the increase in the efficacy and technical participation in the game.

6.8   Conclusions

Coaches may use different task conditions to improve tactical thinking on the players or to increase specific skills during a game (Davids et al. 2013). These conditions may change the game dynamics but using SSCGs it is always possible to keep the main characteristics: teammates, opponents, ball, defensive and attacking actions. For that reason, this chapter aimed to analyze the effects of different conditions in acute physiological responses, time–motion profile and technical performance.

The main conclusions from this chapter are that small goals or endline, limited number of touches on the ball, man-to-man marking, use of floaters, coaches' encouragement, and continuous regimen of training contribute for an increase in acute physiological responses and greater distances covered. Moreover, games without regular goals, man-to-man marking and with floaters may contribute for an increase in skills performed during games and also to an improvement in tactical behavior and collective organization.

These indicators may be used to design the SSCGs and to identify the best periods of the week to apply. The distribution of SSCGs by the weekly periodization seems to be the ultimate objective of coaches and for such reason the following chapter will try to summarize the main evidences verified and provide some examples and guidelines to prescribe SSCGs during a typical week of training.

References

Aguiar, M., Botelho, G., Lago, C., Maças, V., & Sampaio, J. (2012). A review on the effects of soccer small-sided games. *Journal of Human Kinetics, 33*, 103–113.

Almeida, C. H., Ferreira, A. P., & Volossovitch, A. (2012). Manipulating task constraints in small-sided soccer games: Performance analysis and practical implications. *The Open Sports Sciences Journal, 5*, 174–180.

Aroso, J., Rebelo, A. N., & Gomes-Pereira, J. (2004). Physiological impact of selected game-related exercises. *Journal of Sports Sciences, 22*, 522.

Bekris, E., Gissis, I., Sambanis, M., Milonys, E., Sarakinos, A., & Anagnostakos, K. (2012). The physiological and technical-tactical effects of an additional soccer player's participation in small sided games training. *Physical Training, Oct 2012*.

Casamichana, D. G., Castellano Paulis, J., González-Morán, A., García-Cueto, H., & García-López, J. (2011). Demanda fisiológica en juegos reducidos de fútbol con diferente orientación del espacio (Physiological demand in small-sided soccer games with different orientation in space). *Revista Internacional de Ciencias Del Deporte, 7*, 141–154.

Casamichana, D., Castellano, J., Blanco-Villaseñor, Á., & Usabiaga, O. (2012). Study of perceived exertion in soccer training tasks with the generalizability theory. *Revista de Psicología Del Deporte, 21*(1), 35–40.

Casamichana, D., Castellano, J., & Dellal, A. (2013a). Influence of different training regimes on physical and physiological demands during small-sided soccer games. *Journal of Strength and Conditioning Research, 27*(3), 690–697. doi:10.1519/JSC.0b013e31825d99dc.

Casamichana, D., San Román-Quintana, J., Calleja-González, J., & Castellano, J. (2013b). Use of limiting the number of touches of the ball in soccer training: Does it affect the physical and physiological demands? *RICYDE. Revista Internacional de Ciencias Del Deporte, 9*(33), 208–221. doi:10.5232/ricyde2013.03301.

Casamichana, D., Suarez-Arrones, L., Castellano, J., & Román-Quintana, J. S. (2014). Effect of number of touches and exercise duration on the kinematic profile and heart rate response during small-sided games in soccer. *Journal of Human Kinetics, 41*(1), 113–123. doi:10.2478/hukin-2014-0039.

Casamichana, D., Román-Quintana, J. S., Castellano, J., & Calleja-González, J. (2015). Influence of the type of marking and the number of players on physiological and physical demands during sided games in soccer. *Journal of Human Kinetics, 47*(1), doi:10.1515/hukin-2015-0081.

Cihan, H. (2015). The effects of defensive strategies on the physiological responses and time-motion characteristics in small-sided games. *Kinesiology, 47*(2), 179–187.

Clemente, F. M., Martins, F. M., & Mendes, R. S. (2014a). Developing aerobic and anaerobic fitness using small-sided soccer games: methodological proposals. *Strength and Conditioning Journal, 36*(3), 76–87.

Clemente, F. M., Wong, D. P., Martins, F. M. L., & Mendes, R. S. (2014b). Acute effects of the number of players and scoring method on physiological, physical, and technical performance in small-sided soccer games. *Research in Sports Medicine (Print), 22*(4), 380–397. doi:10.1080/15438627.2014.951761.

Clemente, F. M., Martins, F. M. L., Mendes, R. S., & Campos, F. (2015). Inspecting the performance of neutral players in different small-sided games. *Motriz: Revista de Educação Física, 21*(1), 45–53. doi:10.1590/S1980-65742015000100006.

Davids, K., Araújo, D., Correia, V., & Vilar, L. (2013). How small-sided and conditioned games enhance acquisition of movement and decision-making skills. *Exercise and Sport Sciences Reviews, 41*(3), 154–161.

Dellal, A., Chamari, K., Owen, A. L., Wong, D. P., Lago-Penas, C., & Hill-Haas, S. (2011). Influence of technical instructions on the physiological and physical demands of small-sided soccer games. *European Journal of Sport Science, 11*(5), 341–346. doi:10.1080/17461391.2010.521584.

Duarte, R., Araújo, D., Fernandes, O., Travassos, B., Folgado, H., Diniz, A., & Davids, K. (2010). Effects of different practice task constraints on fluctuations of player heart rate in small-sided football games. *The Open Sports Sciences Journal, 3*, 13–15.

Evangelos, B., Eleftherios, M., Aris, S., Ioannis, G., Konstantinos, A., & Natalia, K. (2012). Supernumerary in small sided games 3vs3 & 4vs4. *Journal of Physical Education and Sport, 12*(3), 398–406.

Fanchini, M., Azzalin, A., Castagna, C., Schena, F., McCall, A., & Impellizzeri, F. M. (2011). Effect of bout duration on exercise intensity and technical performance of small-sided games in soccer. *Journal of Strength and Conditioning Research/National Strength & Conditioning Association, 25*(2), 453–458. doi:10.1519/JSC.0b013e3181c1f8a2.

Halouani, J., Chtourou, H., Dellal, A., Chaouachi, A., & Chamari, K. (2014a). Physiological responses according to rules changes during 3 vs. 3 small-sided games in youth soccer players: stop-ball vs. small-goals rules. *Journal of Sports Sciences*, (April), 37–41. doi:10.1080/02640414.2014.899707.

Halouani, J., Chtourou, H., Gabbett, T., Chaouachi, A., & Chamari, K. (2014b). Small-sided games in team sports training: A brief review. *The Journal of Strength and Conditioning Research, 28*(12), 3594–3618.

Hill-Haas, S. V., Rowsell, G. J., Dawson, B. T., & Coutts, A. J. (2009). Acute physiological responses and time-motion characteristics of two small-sided training regimes in youth soccer players. *Journal of Strength and Conditioning Research, 23*(1), 111–115.

Hill-Haas, S. V., Coutts, A. J., Dawson, B. T., & Rowsell, G. J. (2010). Time-motion characteristics and physiological responses of small-sided games in elite youth players: the influence of player number and rule changes. *Journal of Strength and Conditioning Research, 24*(8), 2149–2156. doi:10.1519/JSC.0b013e3181af5265.

Hill-Haas, S. V., Dawson, B., Impellizzeri, F. M., & Coutts, A. J. (2011). Physiology of small-sided games training in football. *Sports Medicine, 41*(3), 199–220.

Köklü, Y. (2012). A comparison of physiological responses to various intermittent and continuous small-sided games in young soccer players. *Journal of Human Kinetics, 31*, 89–96.

Mallo, J., & Navarro, E. (2008). Physical load imposed on soccer players during small-sided training games. *The Journal of Sports Medicine and Physical Fitness, 48*, 166–171.

Ngo, J. K., Tsui, M. C., Smith, A. W., Carling, C., Chan, G. S., & Wong, D. P. (2012). The effects of man-marking on work intensity in small-sided soccer games. *Journal of Sports Science & Medicine, 11*(1), 109.

Rampinini, E., Impellizzeri, F. M., Castagna, C., Abt, G., Chamari, K., Sassi, A., & Marcora, S. M. (2007). Factors influencing physiological responses to small-sided soccer games. *Journal of Sports Sciences, 25*(6), 659–666.

Renshaw, I., Araújo, D., Button, C., Chow, J. Y., Davids, K., & Moy, B. (2015). Why the constraints-led approach is not teaching games for understanding: a clarification. *Physical Education and Sport Pedagogy*, 1–22. doi:10.1080/17408989.2015.1095870.

Román-Quintana, J. S., Casamichana, D., Castellano, J., Calleja-Gonzalez, J., Jukic, I., & Ostojic, S. (2013). The influence of ball-touches number on physical and physiological demands of large-sided games. *Kinesiology, 45*(2), 171–178.

Sampaio, J., Garcia, G., Macas, V., Ibanez, J., Abrantes, C., & Caixinha, P. (2007). Heart rate and perceptual responses to 2 x 2 and 3 x 3 small-sided youth soccer games. *Journal of Sports Science & Medicine, 6*(10), 121–122.

Sampaio, J. E., Lago, C., Gonçalves, B., Maçãs, V. M., & Leite, N. (2014). Effects of pacing, status and unbalance in time motion variables, heart rate and tactical behaviour when playing 5-a-side football small-sided games. *Journal of Science and Medicine in Sport, 17*(2), 229–233. doi:10.1016/j.jsams.2013.04.005.

Chapter 7
Periodization of Training Based on Small-Sided and Conditioned Games

Abstract The knowledge about the acute effects and adaptations that result from the small-sided and conditioned games (SSCG) may help the training periodization, particularly considering the weekly microperiodization. This chapter will summarize the physiological, physical, technical, and tactical effects of different task conditions. After that, a set of methodological considerations will be provided to help the coaches to identify the best periods of the week to apply specific SSCG. Finally, a proposal of weekly microcycle will be provided trying to help the coaches to identify the applicability of the scientific findings about these games in the practical context of the training.

Keywords Periodization · Small-sided and conditioned games · SSG · Soccer · Football · Training methodology

7.1 Introduction

The acute effects of each task condition in physiological, physical and technical performance were described in the previous chapters. After that, it is now moment to summarize the main evidences and highlight the most appropriate games for each period of the week. A brief graphical representation of the causes to increase and decrease acute physiological responses during small-sided and conditioned games (SSCG) can be observed in the following Fig. 7.1.

Moreover, using the information collected from different studies, a graphical representation was also generated to demonstrate the effects of task conditions in technical performance (Fig. 7.2). Both Figs. 7.1 and 7.2 aims to quickly identify the effects of different task conditions and which of these conditions must be used to design new SSCGs in the daily work of coaches.

If the aim is to maximize the intensity of exercise, smaller formats (1 vs. 1–4 vs. 4) and bigger fields contribute for the first step of SSCGs design. After that, coach may use touches limitations (1 to 3 touches as limit), no goals, man-to-man marking, and encouragement during the task. These variables contribute to reach heart rate values

© The Author(s) 2016
F.M. Clemente, *Small-Sided and Conditioned Games in Soccer Training*,
SpringerBriefs in Applied Sciences and Technology,
DOI 10.1007/978-981-10-0880-1_7

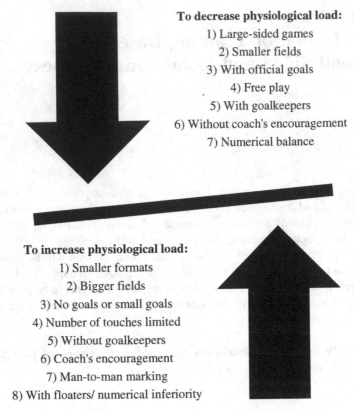

To decrease physiological load:
1) Large-sided games
2) Smaller fields
3) With official goals
4) Free play
5) With goalkeepers
6) Without coach's encouragement
7) Numerical balance

To increase physiological load:
1) Smaller formats
2) Bigger fields
3) No goals or small goals
4) Number of touches limited
5) Without goalkeepers
6) Coach's encouragement
7) Man-to-man marking
8) With floaters/ numerical inferiority

Fig. 7.1 Increasing and decreasing acute physiological responses using different task conditions in SSCGs

and blood lactate concentrations in the anaerobic levels. Nevertheless, the intensity of each task must be prescribed using intermittent regimen of training. Moreover, it is also important to highlight that coach may not use all the variables in the same game. The manipulation of variables must carefully respect the tactical aim of task and the plan for the training. The task should follow the training sequence and organization, and for that reason some conditions may not make sense in some training sessions or for some organizations.

Not only physiological and physical development must be considered to design the game. Technical performance must also be an important variable to be considered by coach. Smaller fields and smaller formats increase the individual participation of each player during the game. Nevertheless, such increase in the participation may also raise the fatigue effects. In that moment, coach must consider what he wants from the task: (1) develop technical performance; or (2) develop fitness levels. If the option is (1), it is important to increase the time to rest and decrease the intensity of exercise. Without such decrease in intensity, technical performance can be compromised. A reduction in motor coordination is normally

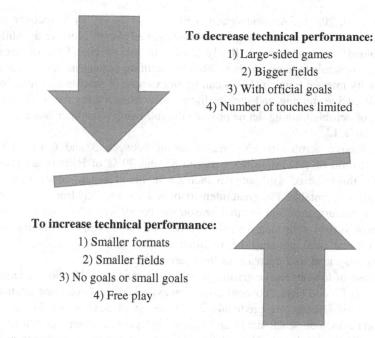

To decrease technical performance:
1) Large-sided games
2) Bigger fields
3) With official goals
4) Number of touches limited

To increase technical performance:
1) Smaller formats
2) Smaller fields
3) No goals or small goals
4) Free play

Fig. 7.2 Increasing and decreasing technical performance using different task conditions in SSCGs

verified in levels above 6 mmol/L of blood lactate concentrations (Janssen 2001). On the other hand, if the aim is to develop fitness levels, coach may follow a regular protocol for fitness development.

7.2 Aerobic and Anaerobic Development

The acute physiological effects of SSCGs lead to the intensities for aerobic and aerobic workout. For that reason, it is only necessary to correctly prescribe the games based on the knowledge of training methodology. Both aerobic and anaerobic systems must be developed and for that reason our aim is to briefly propose some methodological recommendations for the use of SSCGs considering the target of exercise.

7.2.1 Aerobic Training

Oxygen transport system can be improved by exercise. Tasks with medium- to long durations at submaximal levels are more appropriate to improve aerobic system

(Clemente et al. 2014b). Aerobic system will increase the ability to recover from great efforts, to more resistance at great intensities, and also to promote the shift to right in blood lactate threshold (Reilly 2007). In the specific field of aerobic training, it is possible to consider short-intensive training or long-intensive training. High-intensity interval training (HIIT) can be an example of short-intensive training and long-intensive training and can be linked to continuous regimens. Based on both types of aerobic training, let us provide the following recommendations in the Figs. 7.3 and 7.4.

Short-intensive aerobic training has duration between 3 and 6 min and a work-to-rest ratio of 1:1. Intensities between 85 and 90 % of HRmax are recommended for this type of workout. An increase in blood lactate concentration to 5–8 mmol/L is acceptable. The great intensity of this training may lead to a decrease in technical performance and for that reason the complexity of exercise must be small in comparison with long-intensive aerobic training. Simple skills and not a complex tactical thinking must be required on this kind of exercise. A great complexity may lead to a decrease in the exercise's intensity.

In the case of long-intensive aerobic training, a period of 6–15 min is adequate. Small values of blood lactate concentrations are expectable and heart rate intensities below 90 % of HRmax are predictable. The design of SSCGs for this kind of workout must include small fields and large-sided games. Other conditions recommended for this kind of training are free play, no coach's encouragement, and no floaters. A great tactical complexity must be introduced. The coach must emphasize

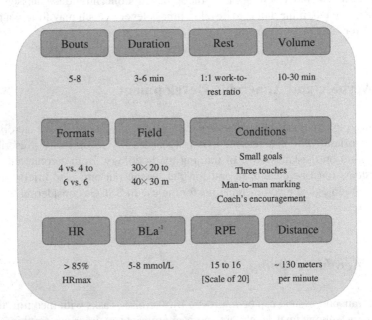

Fig. 7.3 Recommendations to design SSCGs for short-intensive aerobic training

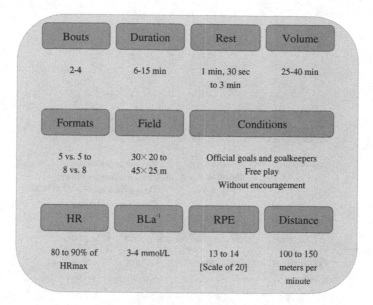

Fig. 7.4 Recommendations to design SSCGs for long-intensive aerobic training

tactical principles and tactical behaviors during this period. For that reason, less encouragement and more valuable and pertinent feedback are desirable.

7.2.2 Anaerobic Training

Glycolytic system is often present during soccer games. Values closer to 12 mmol/L have been found in elite soccer players, thus suggesting that lactate-producing energy system is highly stimulated. For that reason, the ability to repeatedly perform high-intensity tasks must be developed using anaerobic training (Clemente et al. 2014b; Reilly 2007). Smaller formats of the game and bigger fields are recommended to increase the acute physiological responses. Work-to-rest ratios of 1:1 or 1:1.5 are recommended for very high workouts. Task conditions such as touches limitations, no goals, man-to-man marking, coach's encouragement, and floaters can be recommended. Moreover, a small level of tactical thinking is desirable to keep very high efforts. Technical performance must be more relevant in these tasks than tactical behavior. The following Fig. 7.5 shows the summary of recommendations to design SSCGs for anaerobic training.

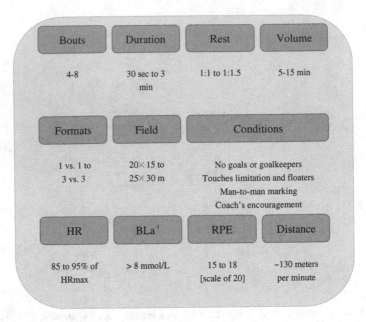

Fig. 7.5 Recommendations to design SSCGs for anaerobic training

7.3 Weekly Training Microcycle: Recommendations

Different periodization strategies may be adopted by coaches. Nevertheless, our aim did not describe the application of SSCGs for many possibilities of weekly training periodization. This option must be integrated in the plan of each coach and considering the requirements and properties of each team. This section will only present an example of weekly periodization for an example of tactical periodization (Delgado-Bordanau and Mendez-Villanueva 2012). Briefly, this periodization ensures stability in the blocks of training and fitness development, and only changes the type of games that is used to develop such capabilities. The main argument is that, during a season, a soccer team must be in a great level in all moments and not in extraordinary levels in some points and in a lower level in others. For that reason, there are very small variations of the load during the season and the weekly training load is almost similar week by week. This argument can be easily discussed by other ideas of periodization; nevertheless our aim is not followed for such route of discussion.

One of the main coaches that use SSCGs as principal tasks to develop his teams is José Mourinho. This coach has won two UEFA Champions Leagues, one UEFA Cup, eight Championships in four different countries (Portugal, England, Italy and Spain) with clubs such as Real Madrid, Chelsea, Inter Milan, and FC Porto. During a long interview about their training methodologies, with regard to periodization José Mourinho said (Oliveira et al. 2006):

From the second microcycle of the preseason, all microcycles are basically the same until the end of the season. The principles of play and the work targets, as well as the physical training, follows the same microcycle over the season. Only with regard to specific tactical behavior there are some changes based on the opponents. Nevertheless, just talking about physical dimension that are more associated with traditional periodization, the targets are the same from the second until the last microcycle. The first microcycle is just to promote a functional adaptation after vacations, just trying adjust the players to making an effort, nothing more. In this first week we do not aim for any physical increments, but only a specific adaptation to the game. From the second week there are weekly cycles that repeat. Thus, I only use microcycles. My guidelines for the weekly standard are equal in July or April.

As Mourinho emphasizes, first weeks of preseason are used to increase the aerobic levels to support high intensities and efforts during the remaining weeks. This particular case must be considered and for that reason we will try to identify some topics and orientations about this specific moment of the season.

7.3.1 Preseason

Increase the volume and ensure that light to moderate levels of intensity takes the highest priority during the preseason. For that reason, aerobic workouts of moderate intensities are performed regularly during this period of the season (Bangsbo 1994). With the progression of the weeks, aerobic training with low to moderate intensities is gradually replaced by high-intensity workouts (Clemente et al. 2014a). Anaerobic training can also be gradually introduced during the weeks of preseason, fundamentally to improve the recovery capability from high-intensity workouts with great levels of blood lactate production (Clemente et al. 2014a).

Large-sided games (7 vs. 7 to 10 vs. 10) played at smaller fields during 4–5 bouts of 5–15 min and a recovery period of 1–3 min for an overall volume of 30 min is recommended (Clemente et al. 2014a). Gradually, smaller formats (4 vs. 4 to 6 vs. 6) may replace the large-sided games to progressively increase the intensities of training and promote cardiovascular adaptations. Let us provide in the Fig. 7.6, a suggestion of workout priorities during a season considering the suggestions made by Clemente et al. (2014a) in the article Periodization based on small-sided soccer games: theoretical considerations.

7.3.2 Weekly Periodization During the Season

Aerobic endurance performance of soccer players do not significantly change during the in-season period (McMillan et al. 2005). For that reason, after a period of great development (12 weeks after the beginning of preseason) in aerobic levels, the workout of this capability may only be performed during 2–3 sessions of the week. Such workout must also consider the competitive schedule of the team. For the case

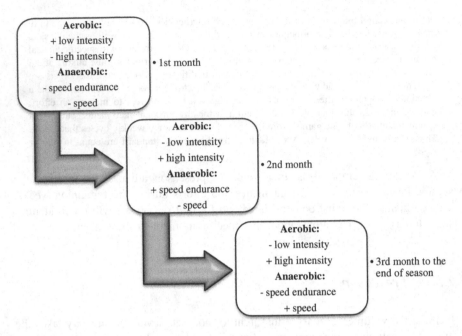

Fig. 7.6 Priority of fitness development during a soccer season

of teams that compete twice a week, the periods between games is only to recover and not properly to workout the capabilities.

Some studies have been showing that the highest volume of training is performed in middle week (Coutinho et al. 2015; Impellizzeri et al. 2004). Postmatch sessions are associated with low intensities (50–70 % HRmax) and the intensities are progressively increased to 70–90 % (middle week) and 90–100 % in the final sessions before the match (Whyte 2006). The preseason sessions are dedicated to short sprints and lower distances covered in comparison with middle-week sessions (Coutinho et al. 2015).

In some weekly periodization, the full rest day occurs in the day after match and in other occurs in the day after postmatch session (Owen and Wong 2009). This option varies from country to country and should be understood in the aim of the weekly workout. Taking into account the Italian example of periodization, there is no rest day in the middle of the week but rather 2 days of tapering are used with low stimulation before the match (Clemente et al. 2014a; Impellizzeri et al. 2004).

The weekly workload should change from team to team considering the fitness levels of players, tactical principles, and model of the game. The games used per session should also respect the main model of play and for that reason are discouraged the use of standard games. A careful analysis to the technical levels, tactical behaviors, and players' potential must be made before the design and application of SSCGs.

Table 7.1 Proposal of a weekly training periodization based on SSCGs

Volume	Match	1 day after match	2 days after match	3 days after match	4 days after match	5 days after match	6 days after match
		Match → Rest	Recovery	Aerobic and strength	Aerobic	Speed	Recovery
Formats			7 vs. 7 to 8 vs. 8	2 vs. 2 to 4 vs. 4	5 vs. 5 to 7 vs. 7	1 vs. 1 to 2 vs. 3	7 vs. 7 to 8 vs. 8
Field (m)			40×25 to 45×25	25×20 to 30×35	30×20 to 40×25	20×15 to 25×30	40×25 to 45×25
Bouts			2 to 4	3 to 6	2 to 4	4 to 8	2 to 4
Duration (min)			6 to 15	3 to 6	6 to 15	0.5 to 3	6 to 15
Work-to-rest ratio			3	01:01	1 to 3	1:1 to 1:1,5	3
Volume (min)			15 to 20	10 to 30	25 to 40	5 to 15	15 to 20
%HRmax	Match	Day off	60 to 75	> 85	80 to 90	85 to 90	60 to 75
Blood lactate (mmol/L)			3 to 4	5 to 8	3 to 4	6 to 8	3 to 4
RPE			13 to 14	15 to 16	13 to 14	15 to 18	13 to 14
Goals			Yes	Yes	No	No	Yes
Floaters			No	No	Yes	Yes	No
Encouragement			No	No	Yes	Yes	No
Man-to-man marking			No	No	Yes	Yes	No

Recovery days (postmatch and prematch) may be an appropriate moment to develop some general tactical principles, mainly considering the large-sided games that should be used in this period of week. Volume and the intensity are low and the training sessions must focus in some tactical issues.

The acquisition days occurs in middle week (3rd, 4th, and 5th days after match). Aerobic and anaerobic systems may be developed in this period of the week, mainly using greater volumes and intensities than in recovery days. Smaller formats and the use of other task conditions must be used to workout specific tactical behaviors and skills.

A proposal of weekly periodization for elite to U19 players is provided in the following Table 7.1. This proposal only makes some general recommendations for a weekly periodization with only one match per week. The design of the tasks must consider the strategies of coaches and the collective organization that should be the main priority to reply the reality of model of play.

7.4 Conclusions

This book was written for designing SSCGs and not to provide a 'magical formula'. A review of the majority of studies conducted in SSCGs was carried out and the fundamental results of these studies were summarized throughout the chapters. Comparison between the effects of SSCGs and traditional running methods was performed. No statistical differences were found. Acute physiological effects of different task conditions were analyzed. The physiological and physical load is similar to running-training methods. Moreover, technical development and tactical behaviors can be also workout during SSCGs. After that, a proposal for a weekly periodization was provided.

Future researches must consider analyzing the effects of different weekly periodization in soccer players. The scientific analysis to this application is small and without such information it will be harder to provide relevant information for coaches and for practical applications. For that reason, this analysis must be the next concerns of researchers interested in SSCGs.

Our aim was to provide the available information about SSCGs and to emphasize the benefits of SSCGs for soccer training. Nevertheless, the use of such games must be managed based on the principles of play and the orientations of the coach. For that reason, it is not recommended to provide examples of games. Provide scientific data and information that is more relevant to give the opportunity to coaches to design their own drills. This follows one old proverb: 'give a man a fish and you feed him for a day; teach a man to fish and you feed him for a lifetime'.

References

Bangsbo, J. (1994). *Fitness training in football—a scientific approach.* Bagsværd, Denmark: HO +Storm.

Clemente, F. M., Martins, F. M. L., & Mendes, R. S. (2014a). Periodization based on small-sided soccer games. *Strength and Conditioning Journal, 36*(5), 34–43.

Clemente, F. M., Martins, F. M., & Mendes, R. S. (2014b). Developing aerobic and anaerobic fitness using small-sided soccer games: Methodological proposals. *Strength and Conditioning Journal, 36*(3), 76–87.

Coutinho, D., Gonçalves, B., Figueira, B., Abade, E., Marcelino, R., & Sampaio, J. (2015). Typical weekly workload of under 15, under 17, and under 19 elite Portuguese football players. *Journal of Sports Sciences, 33*(12), 1229–1237.

Delgado-Bordanau, J. L., & Mendez-Villanueva, A. (2012). Tactical periodization: Mourinho's best-kept secret? *Soccer Journal,* 29–34.

Impellizzeri, F. M., Rampinini, E., Coutts, A. J., Sassi, A., & Marcora, S. M. (2004). Use of RPE-based training load in soccer. *Medicine and Science in Sports and Exercise, 36*(6), 1042–1047.

Janssen, P. (2001). *Lactate threshold training.* Champaing, IL: Human Kinetics.

McMillan, K., Helgerud, J., Grant, S., Newell, J., Wilson, J., Macdonald, R., & Hoff, J. (2005). Lactate threshold responses to a season of professional British youth soccer. *British Journal of Sports Medicine, 39,* 432–436.

Oliveira, B., Amieiro, N., Resende, N., & Barreto, R. (2006). *Mourinho: Porquê tantas vitórias? [Mourinho: Why so many victories?].* Lisboa, Portugal: Gradiva.

Owen, A. L., & Wong, P. L. (2009). In-season weekly high-intensity training volume among professional English soccer players: A 20-week study. *Soccer Journal,* 28–32.

Reilly, T. (2007). *The science of training—soccer.* Oxon, UK: Routledge.

Whyte, G. (2006). *The physiology of training.* London, UK: Churchill Livingstone Elsevier.

Printed in the United States
By Bookmasters